DATE DUE

FE 13 03			

DEMCO 38-296

Oryx Frontiers of Science Series
CHEMISTRY

by David E. Newton

ORYX PRESS
1999

*The rare Arabian Oryx is believed to have inspired the myth of the unicorn.
This desert antelope became virtually extinct in the early 1960s. At that time,
several groups of international conservationists arranged to have nine
animals sent to the Phoenix Zoo to be the nucleus of a captive breeding
herd. Today, the Oryx population is over 1,000, and over 500 have
been returned to the Middle East.*

© 1999 by David E. Newton
Published by The Oryx Press
4041 North Central at Indian School Road
Phoenix, Arizona 85012-3397

Cover image supplied by Dr. Tahir Cagin of California Institute of Technology. The
image represents a snapshot from Molecular Dynamics Simulation of a generation 6
PAMAM dendrimer with ethylenediamine core. The simulation was carried out at the
Materials and Process Simulation Center by T. Cagin and W. A. Goddard, III.

Published simultaneously in Canada
Printed and bound in the United States of America

∞ The paper used in this publication meets the minimum requirements of
American National Standard for Information Science—Permanence
of Paper for Printed Library Materials, ANSI Z39.48, 1984.

Library of Congress Cataloging-in-Publication Data

Newton, David E.
    Chemistry / by David E. Newton
      (Oryx frontiers of science series)
    Includes bibliographical references and index.
    ISBN 1-57356-160-6
    1. Chemistry. I. Title. II. Series.
    QD31.2.N49 1999
    540—dc21   98-40880
                                                     CIP

For Rick Kimball,
"my editor" for those many years.

Your professional, community, and personal works
have long been a great inspiration
to me and many others

CONTENTS

Preface   vii

Chapter 1. Chemistry Today   1
Chapter 2. Chemical Technology Today   72
Chapter 3. Social Issues in Chemistry   116
Chapter 4. Biographical Sketches   135
Chapter 5. Documents and Reports   163
Chapter 6. The Future of Chemistry   196
Chapter 7. Career Information   204
Chapter 8. Statistics and Data   221
Chapter 9. Organizations and Associations   235
Chapter 10. Print and Electronic Resources   259
Chapter 11. Glossary   275

Index   285

PREFACE

hese are exciting times in the chemical sciences. Scarcely a week goes by without some significant new discovery being announced. A researcher may report the development of a molecule that can reproduce itself; or a substance that can change from transparent to opaque at different pH values; or a new kind of transistor with 10 times the speed of the best material now available. The task of staying abreast of such changes across the entire field of the chemical sciences is probably more than most professionals are able to manage. It is an even more improbable task for the nonspecialist.

The purpose of this book is to provide readers with a flavor of the kinds of advances now being made in the chemical sciences. The author has chosen to focus on developments announced within the last three years (1996–1998). This time frame, although arbitrary, makes it possible to concentrate on a select number of recent and important developments in the fields of chemistry and chemical engineering. The criterion for selecting these topics was that they be mentioned and discussed in at least one journal or magazine designed for the general reader interested in chemistry. Such journals and magazines include *Science News*, *Scientific American*, and *Chemistry & Industry*.

Chapters 1 and 2 of the book summarize about two dozen research studies in chemistry and chemical engineering. Some of these studies are relatively simple and fairly concrete, while others are more complex and abstract. The author has assumed that readers will have completed at least one year of high school chemistry and are conversant with most fundamental concepts taught at that level.

Readers with a general interest in the subject of chemistry should find a ready supply of topics to satisfy their interest. For those wishing to continue their research on the topics presented here, such as high school and college chemistry students, background reading materials are suggested for further study. Over time, many of these discoveries are likely to become "landmark" accomplishments that will lead to other important findings. This volume may be able to serve, therefore, as a reference in the future with regard to the origin of important trends in the field of chemistry.

The book also contains chapters on a number of other topics related to developments in the chemical sciences. Chapter 3 discusses some of the social, ethical, political, and philosophical issues associated with changes in the chemical sciences. For example, parts of chapter 3 discuss how advances in chemistry have led to evolving issues with regard to the Earth's atmosphere and to ways by which those issues can be resolved.

Other chapters of the book deal with other topics of interest to readers who want to know more about the chemical enterprise and how it goes forward. For example, chapter 4 provides biographical sketches of those researchers whose work is mentioned in chapters 1 and 2. In this chapter, readers can see the variety of ways in which individuals come to chemistry, the settings in which they work, and the progress their careers may follow.

Chapter 5 is a resource chapter. It contains a number of documents relating to the socioscientific issues described in chapter 3.

Chapter 6 provides readers with a glimpse of the future of chemistry and chemical engineering. What problems will chemists of the next 25 years be working on? What kinds of challenges will they face? What kinds of breakthroughs might they be expected to make? The reports of two major committees in chemistry and chemical engineering are reviewed in chapter 6 in an effort to find answers to these questions.

Chapter 7 is devoted to a review of career information about chemistry and chemical engineering. Some of the kinds of jobs available in these fields are described, and further resources on careers in chemistry are reviewed.

Chapter 8 is a collection of a relatively wide variety of statistics and data. Readers should be able to develop a better feeling for the direction that chemistry is taking by reviewing trends in production, employment, financing, and other aspects of the chemical sciences.

Chapter 9 provides a list of organizations and associations for those working or interested in the fields of chemistry and chemical engineering.

Such associations are often gold mines of information for those who want to know more about topics such as advances in some given field in chemistry, job and training opportunities, meetings and classes, and political issues and political action.

Chapter 10 summarizes some major print and electronic resources in the chemical sciences. In keeping with the book's focus on the last three years of research, this chapter focuses on journals, magazines, and Internet sources. Books are not included in this chapter since most of the discoveries discussed in chapters 1 and 2 have not yet had time to find their way into books that are accessible to the general reader.

Finally, chapter 11 provides a glossary of terms. This glossary does not include terms generally introduced in a first year chemistry class. It does include more specialized terms that arise in chapters 1, 2, and 3.

CHAPTER ONE
Chemistry Today

W hat are the most interesting breakthroughs in chemical research that have occurred over the past three years? Not so very long ago, that question might have been much easier to answer. The pace of chemical research, while not sluggish, was certainly less brisk than it is today. Thus, trying to decide which scientific discoveries are of "special interest" today is a real challenge.

The selections that follow, therefore, should be assumed to be representative of the kinds of progress taking place in chemical research, rather than a list of "the most important." They should provide, however, a taste of some of the directions in which chemical research is headed and the kinds of work on which chemists have been focusing.

The research summarized in chapter 1 covers a wide variety of topics, which have been grouped into five broad categories: Elements, Molecules and Bonding, Chemical Reactions, Biological Molecules, and Biochemical Reactions. These categories have been established to aid the reader, however, and do not necessarily represent common themes among the research. Indeed, other systems of organizing these summaries are possible.

The large number of topics in the last two categories reflects the increasing use of the methods of chemistry to better understand the materials of which living organisms are made and the chemical changes that take place in those materials. It seems likely that this trend in chemistry may become even more pronounced in the future.

ELEMENTS

The research on elements reported below follows two general lines. First, a number of new chemical elements, referred to as transfermium elements, have been discovered. These elements have not been found in the Earth's surface, but have been produced synthetically in particle accelerators. A great deal of research in the last three years has focused on the production and characterization of these elements.

Second, research has produced new information on well-known elements. The discovery of a metallic phase of hydrogen and the ability of sulfur to become superconductive are examples of this line of research.

Transfermium Elements

The transfermium elements are elements with atomic numbers greater than that of fermium (Fm), element #100. None of these elements exists in nature, and nuclear scientists have been working for more than three decades to produce them synthetically in particle accelerators. Of special interest are the elements with atomic numbers greater than 103 (lawrencium, Lr). The reason for this interest is that lawrencium ends the actinide series of elements, which includes elements with atomic numbers from 90 through 103. Elements with atomic numbers of 104 and greater, therefore, are each positioned in a different column, as they belong to different groups. Thus, the question of greatest interest regarding these newly discovered elements has been whether they have properties that correspond to those predicted by the periodic law.

Production of the Transfermium Elements

Answering questions about the chemical and physical properties of these new, super-heavy elements is very difficult because of the problems involved in producing them. The larger the atomic nucleus gets, the more protons there are within the nucleus. The greater the number of protons in the nucleus, the stronger the repulsion of these like-charged particles (the protons) within the nucleus becomes. For this reason, none of the elements beyond bismuth (Bi, #83) has any stable isotopes. The only isotopes known for the actinide elements are radioactive, with half-lives ranging from less than a millisecond to billions of years.

The technique for manufacturing transfermium elements is similar to that used for actinide elements of lower atomic masses. A relatively small particle (such as an alpha particle) is accelerated to high speeds in a particle accelerator. That particle is then directed at a target atom. Under the proper circumstances, the small particle collides and fuses with the larger atom, forming a new atom. Thus, the element curium (Cm, #96) can be made by bombarding a plutonium target with alpha particles:

$$^{239}_{94}\text{Pu} + ^{4}_{2}\text{He} \rightarrow ^{243}_{96}\text{Cm}$$

Curium-243 has a half-life of 32 years and decays via the emission of an alpha particle.

In theory, one could manufacture very large atoms, such as an atom with atomic number 114, by bombarding an atom with atomic number 112 with an alpha particle, as in the method described above. In fact, this is not an option. The target atom in this case (atomic number 112) decays so rapidly that it does not remain in existence long enough to be struck by an alpha particle.

The alternative method for making super-heavy atoms is by using larger particles in the bombardment process. For example, element 104 was first produced in the 1960s by bombarding plutonium-242 with ions of neon-22:

$$^{242}_{94}\text{Pu} + ^{22}_{10}\text{Ne} \rightarrow ^{264}104$$

Notice that no chemical symbol is provided for element 104 in the above equation. The reason for this omission is that the name given to newer elements is often a matter of considerable dispute, resolved only after many years or even decades. The problem of naming the transfermium elements is discussed in more detail later.

Some of the mechanisms by which elements beyond #104 have been prepared are summarized below.

$$^{243}_{95}\text{Am} + ^{22}_{10}\text{Ne} \rightarrow ^{265}105$$

$$^{249}_{98}\text{Cf} + ^{18}_{8}\text{O} \rightarrow ^{267}106$$

$$^{209}_{83}\text{Bi} + ^{54}_{24}\text{Cr} \rightarrow ^{263}107$$

$$^{208}_{82}\text{Pb} + ^{58}_{26}\text{Fe} \rightarrow ^{266}108$$

$$^{209}_{83}\text{Bi} + ^{58}_{26}\text{Fe} \rightarrow ^{267}109$$

$$^{208}_{82}\text{Pb} + ^{62}_{28}\text{Ni} \rightarrow ^{270}110$$

$$^{209}_{83}\text{Bi} + ^{62}_{28}\text{Ni} \rightarrow ^{271}111$$

$$^{208}_{82}\text{Pb} + ^{69}_{30}\text{Zn} \rightarrow ^{277}112$$

(Note: Other methods of production are possible for all of these elements.)

Credit for Discovery

Research on the transfermium elements requires very large and expensive machinery. In fact, such research has been conducted at only three laboratories throughout the world: the Joint Institute of Nuclear Research, at Dubna, Russia; the Lawrence Berkeley Laboratory at the University of California at Berkeley in the United States; and the Gesellschaft für Schwerionenforschung (Institute for Heavy Ion Research, or GSI) in Darmstadt, Germany.

In nearly every instance, the discovery of a new heavy element has been announced by at least two of these three laboratories over a period of time. The reason for multiple discoveries is that identifying the product of any one of the above reactions is very difficult. In most cases, no more than a handful of atoms is produced in any one experiment. Since those atoms all have very short half-lives, scientists must devise ingenious schemes to prove that a new element was actually produced.

In general, identification of a new element is based on the element's decay scheme. Atoms of the element exist for such brief periods of time that only the scantiest of information about their properties can be collected. Suppose, however, that an experiment is conducted that results in the emission of an alpha particle and the formation of a nucleus such as $^{273}110$:

$$X \rightarrow {}^{4}_{2}He + {}^{273}110$$

Working backwards, then, it is evident that X must have had an atomic number of 112 and an atomic mass of 277.

The discovery of element 104 provides an example of the delay that often occurs in recognizing the discovery of a new element. Production of this element was first claimed in 1964 by the Dubna research team. However, few scientists were entirely convinced by the evidence presented by this team. Not until five years later was there sufficient evidence—this time, produced at Berkeley—for scientists to acknowledge that a new element had actually been made.

Nomenclature

Gaining credit for discovering a new element is, of course, important to all three research teams. For one thing, all of the teams would like to receive the recognition that goes with such an accomplishment. At least as important, however, is that, historically, the person or group of people who discover a new element are given the right to name the element.

In the case of element 104, the Russian research team suggested the name *kurchatovium*, in honor of the great Russian nuclear chemist Igor Vasilevich Kurchatov. The American team, on the other hand, proposed the name *rutherfordium* for the new element, in honor of the great British scientist Sir Ernest Rutherford.

Similar situations arose with the discovery of nearly every other transfermium element. At least two names—and sometimes as many as three—were suggested for each of the elements with atomic numbers from 104 to 112.

All final decisions on nomenclature are made by the International Union of Pure and Applied Chemistry (IUPAC). Special committees of IUPAC struggled for more than three decades trying to decide on names for these elements. In 1994, a special IUPAC committee thought it had solved the problem of naming the transfermium elements, and they proposed the following names:

101	Mendelevium (Md)		106	Rutherfordium (Rf)
102	Nobelium (No)		107	Bohrium (Bh)
103	Lawrencium (Lr)		108	Hahnium (Hh)
104	Dubnium (Db)		109	Meitnerium (Mt)
105	Joliotium (Jl)			

The committee's decision produced an uproar. All three research teams thought they had been cheated out of one or more names for which they should have been responsible. Perhaps the greatest debate focused on element 106, which the Berkeley team had proposed naming seaborgium, after Glenn Seaborg, codiscoverer of plutonium and nine other heavy elements. Seaborgium had not been allowed, the IUPAC committee explained, because elements could not be named after living persons (although, both einsteinium [Es] and fermium [Fm] were named while Albert Einstein and Enrico Fermi were still alive).

So many objections were raised that the general assembly of IUPAC rejected its committee's report. Three years later, a new list of names was proposed by the IUPAC Commission on Nomenclature of Inorganic Chemistry and was accepted by the IUPAC Council. Thus, elements 101 through 109 now have the following official names and chemical symbols:

101	Mendelevium (Md)		106	Seaborgium (Sg)
102	Nobelium (No)		107	Bohrium (Bh)
103	Lawrencium (Lr)		108	Hassium (Hs)
104	Rutherfordium (Rf)		109	Meitnerium (Mt)
105	Dubnium (Db)			

Chemical Properties

For chemists, the most interesting questions about the super-heavy elements are those regarding their chemical and physical properties. For example, is rutherfordium (Rf, #104) similar to titanium, zirconium, and hafnium in Group 4 above it, and is dubnium (Db, #105) similar to vanadium, niobium, and tantalum in Group 5 above it?

Answering such questions is extraordinarily difficult because no more than a half dozen or so atoms of each element are produced at any one time. Some of the earliest research on element 106, for example, involved bombarding a 12 cm^2 curium target with a beam of neon atoms emitting 13 trillion atoms per second. After a period of several weeks, researchers had detected just 10 atoms of element 106.

Nonetheless, scientists have developed imaginative and sophisticated techniques for studying even these few short-lived atoms. As it turned out, the results of chemical tests on elements 104 and 105 were both surprising and disappointing. These elements did *not* exhibit the properties expected of them based on their positions within the periodic table. For example, element 105 appeared to behave more like the actinide protactinium, in some instances, than like niobium and tantalum.

The reason for this discrepancy has still not been determined. Some scientists think, however, that relativistic effects may be involved. As the size of an atomic nucleus gets larger, the attraction it exerts on the outermost energy level of electrons greatly increases. This force may distort the energy levels and produce chemical effects not expected or observed in smaller atoms.

A series of experiments carried out in 1997 at Darmstadt on element 106 was, therefore, of very special interest to chemists. As described below, the GSI group found a way to carry out relatively sophisticated tests that required no more than a half-dozen atoms of the elements.

As soon as atoms of element 106 were produced in the particle accelerator, they were transferred to very tiny solid particles in an aerosol, something like the aerosol spray in a deodorant or hair-spray product. The aerosol particles transported the atoms of element 106 through capillary tubes into one of two chromatographic systems. Along the way, the atoms of element 106 were exposed to a variety of chemical agents, such as oxygen, chlorine, and thionyl chloride. Chemists predicted that element 106 would behave like its group members molybdenum (Mo) and tungsten (W), forming compounds such as molybdenum dioxychloride (MoO_2Cl_2) and tungsten dioxychloride (WO_2Cl_2).

One of the chromatographic systems was a gas chromatograph designed to separate molybdenum, tungsten, and seaborgium dioxychlorides from each other, assuming that the last of these compounds was actually present in the mixture. The results showed that separation did occur, and that seaborgium dioxychloride formed at a temperature predicted by its position in the periodic table.

The other chromatographic system involved a liquid separation in which molybdenum, tungsten, and seaborgium were first converted to hexavalent ions and then separated from each other. Again, the separation occurred as expected, with the lightest element (molybdenum) escaping from the chromatographic column first, the next heavier element (tungsten) escaping next, and the heaviest element (seaborgium) escaping last.

Chemists around the world were ecstatic to learn that element 106 had behaved as predicted by the periodic table. The reason it did *not* show erratic behavior like that of elements 104 and 105, however, has as yet to be explained.

The Next Plateau of Stability

Research on transfermium elements has some very profound theoretical implications also. Protons and neutrons in an atomic nucleus occupy specific energy levels similar to those that exist for electrons. Nuclei with "full" energy levels appear to be much more stable than those with incomplete energy levels. The number of protons or neutrons needed to fill an energy level has been referred to as a "magic number." The magic numbers for protons and neutrons are 2, 8, 20, 28, 50, 82, and so on. The "magic numbers" represent that number of protons or neutrons believed needed to fill shells of either particle in the nucleus of an atom. The addition of the next proton or neutron (proton #21 or neutron #29, for example) requires the opening of a new shell. In some cases, atomic nuclei have double magic numbers and are, therefore, particularly stable. Examples of nuclei with double magic numbers are helium-4 (2 protons and 2 neutrons), oxygen-16 (8 protons and 8 neutrons), and lead-208 (82 protons and 126 neutrons).

The first transuranium nucleus to have a double-magic number of protons and neutrons is element 114-298. The nucleus of this element would contain 114 protons and 184 neutrons, which are both magic numbers. Current theories suggest that this nucleus would be unusually stable, perhaps having a half-life as great as a few billion years.

Future research on heavy atoms, then, is aimed not only at finding out how much larger an atom can be made, but also at determining whether stable isotopes of such elements really can and do exist. If such stable isotopes are discovered, scientists will have much more opportunity to study their properties and to learn a great deal more about the fundamental structure of matter in the universe.

References

Browne, Malcolm W., "Scientists Meet Analytical Challenge of an Ephemeral Element," *New York Times*, 8 July 1997, C3.

Collins, Graham P., "Convincing Evidence Seen for Isotopes of Element 100," *Physics Today*, January 1995, 19–20.

"Elemental Upset," *Science News*, 22 October 1994, 271.

Freemantle, Michael, "Heavy-Element Name Saga Ends," *Chemical & Engineering News*, 8 September 1997, 9–10.

Goss Levi, Barbara, "IUPAC/IUPAP Committee Mediates Custody Battle over Heavy Elements," *Physics Today*, November 1993, 20–21.

"How Heavy Can Atomic Nuclei Be?"<http://www.gsi.de/~demo/wunderland/englisch/Kapitel_02.html> (16 September 1997).

Jacoby, Mitch, "Heavy Elements Back on Track," *Chemical & Engineering News*, 7 July 1997, 5.

Koenig, Robert, "GSI Bags Another New Element," *Science*, 1 March 1996, 1231.

Naeye, Robert, "An Island of Stability," *Discover*, August 1994, 22–23.

"New Element no. 110 Discovered," <http://indigo.lib.lsu/edu/lib/chem/display/new_element.html> (16 September 1997).

"Press Release: Element 112 Discovered at GSI," <http://www.gsi.de/z112e.html> (16 September 1997).

Schädel, M., et al., "Chemical Properties of Element 106 (Seaborgium)," *Nature*, 3 July 1997, 56–58.

Stone, Richard, "An Element of Stability," *Science*, 24 October 1997, 571–72.

Metallic Hydrogen

Few elements are less likely to be thought of as "metallic" than is hydrogen. Under normal ambient conditions, hydrogen is a gas that acts as an excellent electrical insulator. That is, it does *not* conduct an electrical current. This condition is entirely understandable since the two electrons in a hydrogen molecule (H_2) are bound tightly to the hydrogen nuclei by a covalent bond. One of the important characteristics of metals is that they contain a swarm of "free" electrons that move easily from one place to another. When an electric potential is applied to a metal, these "free" electrons travel in one direction or the other, producing an electrical current characteristic of an electrical conductor.

Yet, scientists have expected for more than half a century that, under the proper circumstances, hydrogen could become a metal-like conductor. This theory was first suggested in 1935 by the Hungarian-American physicist Eugene Wigner. Wigner suggested that at very high pressures and relatively very low temperatures, hydrogen could be condensed into a solid. Under those conditions, the hydrogen would exist, he theorized, not as molecules but as individual atoms. In such a case, the electron associated with each hydrogen atom might become mobile and might be capable of conducting an electric current.

Wigner's theory proved nearly impossible to test until recently. The problem was the temperatures and pressures needed to attain the conditions upon which his theory was predicated: a mere few thousands of degrees kelvin at about a million atmospheres of pressure.

Experimental Conditions

By the mid-1990s, however, significant technological progress had made it possible to produce the extreme conditions needed to test Wigner's theory. The key experiments were done at the Lawrence Livermore National Laboratory (LLNL), in Livermore, California, under the direction of Dr. William Nellis. The equipment used in these experiments had been developed at LLNL for work on nuclear weapons and for the study of the composition and structure of planets.

The critical piece of equipment used in the experiments is a two-stage light-gas gun, 20 meters in length. The first stage of this gun consists of a long, narrow tube containing hydrogen gas. A charge of up to 3.5 kilograms of gunpowder is ignited at one end of the tube. The explosion drives a piston down the length of the gun to the opposite end, compressing the hydrogen gas. The far end of the tube is constricted, having a shape somewhat like the tip of a pipette.

The second stage of the light-gas gun consists of a much narrower tube connected to the first stage. A small piece of material called the *impactor* is inserted at the juncture of the two tubes. When the compressed gas from the first tube reaches the juncture, the impactor is forced down the second tube with a very high velocity, ranging from 1 to 8 kilometers per second (up to 18,000 miles per hour).

The second stage of the gun ends at the target. The target is a small box made of aluminum metal, which is wrapped in aluminized mylar. The target box contains the sample to be tested, which is maintained at temperatures close to absolute zero. The purpose of the mylar is to reduce heat transfer from the environment to the sample, which would otherwise cause the sample to boil away completely. The target box also contains electrical leads through which the conductivity of the sample within the box can be measured.

When the impactor strikes the target box, it produces a high-pressure shock wave (of about a million atmospheres) that lasts less than a microsecond. Thus, the conditions of the gas within the target box—very high pressure and relatively very low temperature—meet the conditions needed to test Wigner's theory.

Results

In the experiments conducted at LLNL, both hydrogen and deuterium at a temperature of 20 K were used as targets in the gas gun. The two were each kept in a liquid state rather than a solid state for a variety of reasons. First, liquid hydrogen is much easier to produce, store, and work with than is solid hydrogen. Second, the conditions that developed within the target box would have converted solid hydrogen to the liquid phase anyway. Finally, any practical applications the experiment might have had (regarding the composition of planets, for example) would involve the liquid phase of hydrogen rather than its solid phase.

When the gas gun was fired, pressures ranging from 0.9 to 1.8 Mbar (megabars, or millions of atmospheres) and temperatures ranging from 2,000 to 4,000 K were produced within the hydrogen sample. Studies of the electrical conductivity of the hydrogen at these conditions showed the high resistance expected of a gas or liquid at pressures ranging from about 0.9 to 1.4 Mbar. As pressures increased, however, the electrical resistance showed a constant decrease. At pressures greater than 1.4 Mbar, however, quite remarkable results were obtained. Electrical resistance fell to less than 0.001 ohms per centimeter, a value comparable to that of ordinary metals. Currents of 1 A (ampere) were maintained within the sample for periods of up to 200 nanoseconds. These results provided positive evidence for the formation of a metal-like state in the hydrogen sample.

Practical Applications

The results of the LLNL experiment do have practical applications. Planetary scientists believe that the compositions and structures of the larger planets in our solar system are very different from those of the Earth and other terrestrial planets. Jupiter and Saturn, for example, consist primarily of huge amounts of hydrogen and helium. In these planets' outer atmospheres hydrogen and helium are found in their gaseous states.

Pressures at the center of both planets, however, are believed to be enormous, probably 100 Mbar or more. At these pressures, hydrogen might very well be compressed to a liquid phase that exhibits metallic properties similar to those observed in the LLNL experiments. Scientists believe that the motion of this liquid-metallic hydrogen may generate electrical currents through the planets, which, in turn, are responsible for the very large magnetic fields and radio emissions characteristic of both bodies.

References

Johnston, Don, "Lab Team Hits Success with Metallized Hydrogen," <http:// www.llnl.gov/PAO/NewslineStories/Hmetallization.html> (18 September 1997).

"Jumpin' Jupiter! Metallic Hydrogen," <http://www.llnl.gov/str/Nellis.html> (18 September 1997).

"Jupiter: King of Planets," <http://aps.org/apsnews/0696/11475.html> (18 September 1997).

Lipkin, Richard, "Lightest Metal in the Universe," *Science News*, 20 April 1996, 250–51.

Nellis, W. J., M. Ross, and N. C. Holmes, "Temperature Measurements of Shock-Compressed Liquid Hydrogen: Implications for the Interior of Jupiter," *Science*, 1 September 1995, 1249–52.

Nellis, W. J., S. T. Weir, and A. C. Mitchell, "Metallization and Electrical Conductivity of Hydrogen in Jupiter," *Science*, 16 August 1996, 936–38.

Savoye, Craig, "Fact Sheet: Hydrogen Produced in Metal Form for First Time," <http://www.llnl.gov/PAO/NewsRelease/March96/metalhydrofact.html> (18 September 1997).

Sulfur as a Superconductor

One might be hard-pressed to think of an element less likely to be a superconductor than sulfur. The yellow solid is often used as a classic example of a nonmetal, a material unlikely to conduct an electric current at all. Even less likely is that sulfur could conduct an electric current without resistance. Yet, in 1997, a research team consisting of scientists at the Carnegie Institution Geophysical Laboratory and Center for High Pressure Research and at the Institute of High-Pressure Physics at the Russian Academy of Sciences reported that they had observed superconductive properties in sulfur samples exposed to pressures of about 160 GPa (160 gigapascal, or 160×10^9 pascal, or 160 billion pascal).

Pressures of this order of magnitude can now be produced routinely in a device known as a diamond anvil cell. When sulfur is placed within such a cell, and the temperature is reduced to about 10 K and pressures are increased to 157 GPa, evidence of superconductivity in the element is observed. The transition temperature (the temperature at which a substance becomes superconductive) occurs at 17 K when pressures approach 160 GPa.

This result is striking because it is the highest transition temperature yet reported for a pure element. The so-called high-temperature superconductors discovered over the last decade are all ceramics, alloylike substances consisting of complex mixtures of barium, copper, thallium, strontium, yttrium, and other metal oxides. Sulfur's heavier chalcogenide cousins, selenium and tellurium, also demonstrate superconductivity, but at temperatures significantly lower than that observed for sulfur.

Superconductivity was not measured directly in this study for practical reasons. The size of a diamond anvil cell is so small that connecting electrical leads to the sample is impractical. Instead, researchers measured changes in the magnetic susceptibility of the pressurized sulfur, a property dependent upon and reflective of the element's conductivity.

Authors of the study do not hold out hope for practical applications of their discovery, but they do point out the potential value it holds for gaining a better understanding of theoretical concepts relating to superconductivity. It may also aid in the development of more sophisticated tests for superconductivity of small samples under even higher pressures than those used in this experiment.

References

Struzhkin, Viktor V., et al., "Superconductivity at 10–17 K in Compressed Sulphur," *Nature*, 27 November 1997, 382–83.

Wu, C., "Sulfur: Cool, Compact, and Conductive," *Science News*, 29 November 1997, 340.

MOLECULES AND BONDING

One of the most exciting aspects of chemical research is the ability of chemists to constantly invent new forms of matter. Sometimes these "new forms" of matter are simply minor modifications of materials that have been known for many years. The discovery of a new kind of polymer, for example, is always of interest, but seldom represents a breakthrough in our understanding of nature.

Occasionally, however, these discoveries are worthy of the claim that they are "revolutionary." Research on buckyballs, nanotubes, and dendrimers falls into this category. Other research summarized in this section is of special interest in that it reveals new facets of familiar materials. The discoveries of dihydrogen bonding and the structure of cubane are examples of this kind of research.

Buckyballs and Nanotubes

The Nobel Prize for Chemistry in 1995 was awarded to Richard E. Smalley, Robert F. Curl, Jr., and Harold W. Kroto for their discovery of a new allotrope of carbon known as *buckminsterfullerene*. The substance later became more commonly known as a *fullerene*, and the molecules of which it is made are referred to as *buckyballs*. Buckyballs are soccer-ball-shaped collections of 60 carbon atoms joined to each other in pentagons and hexagons. The molecule was named for the famous architect Buckminster Fuller, who made geodesic dome construction popular in the 1950s.

The discovery of fullerenes in 1985 was remarkable enough on its own. It revealed a form of carbon that had never previously been observed. However, the applications that have been discovered for this new form of carbon have been even more staggering. Scientists have discovered ways of manipulating individual molecules of fullerene to construct a host of new structures. That research has produced some of the most exciting news in the field of nanotechnology (the study of matter at the level of individual atoms and molecules).

For example, suppose that a buckyball molecule is cleaved through its center, dividing it into two 30-carbon segments. Then suppose that a ring of 10 carbon atoms is inserted between the 2 segments and the 3 parts are then joined to each other, thus producing another type of fullerene consisting of 70 carbon atoms. This new molecule has the somewhat squashed, spherical shape of a rugby football.

An apparently unlimited variety of fullerenes can be produced by this technique. The molecules produced have virtually any imaginable shape that can be made with some combination of pentagons, hexagons, and heptagons.

Fullerene-Based Nanotubes

One line of research in this field involves an extension of the process by which a 70-carbon fullerene is produced by the splitting and recombination of a 60-carbon fullerene. Imagine not that a single ring of 10 carbon atoms is inserted between the two 30-carbon units, but that two 10-carbon rings are inserted between the two halves . . . or that three, four, or five such rings are inserted between the two halves. The product of this process is a cylindrical-shaped object, which, as more rings are added, becomes a nanotube. In this respect, a nanotube is a long molecule, one-atom thick, whose structure consists of six- or five-carbon rings closed with a semi-spherical cap of such rings at both ends.

Nanotubes were first discovered in 1991 by the Japanese electron microscopist Sumio Iijima. Iijima found the nanotubes in the sooty material produced when carbon is vaporized in an electric arc.

Another way to think of a carbon nanotube is as a *graphene sheet* that has been wrapped around a central axis to form a cylinder. A graphene sheet is a two-dimensional surface made of carbon atoms only. It gets its name from the fact that the common mineral graphite consists of layers of the material stacked on top of each other, having a structure similar to a deck of cards. Normally, the graphene sheets in a piece of graphite have little or no attraction for each other, giving them the ability to slide back and forth over each other. This property accounts for the friability of graphite (i.e., the ease with it can be crumbled) and its consequent common use in "lead" pencils. When one writes with a "lead" pencil, one simply wipes graphene sheets off the pencil "lead" (graphite).

A graphene sheet has a number of interesting properties. For example, a graphene sheet left to itself has a tendency to curl around a central access and close upon itself. Thus, chemists do not have to use sophisticated methods for converting a graphene sheet into a nanotube; the process takes place spontaneously.

Graphene sheets also have two significant physical properties. First, they have the highest tensile strength of any material known. If one could produce a rope made out of carbon nanotubes, it would be from 50 to 100 times stronger than steel, making it the strongest material ever produced. Second, the density of carbon atoms in a graphene sheet is greater than the density of any other two-dimensional material made of a single element. Thus, under normal circumstances, a nanotube or nanorope would be essentially impermeable. When Smalley became aware of these properties of carbon nanotubes, he noted, "That's got to be good for something!"(1996)

Practical Applications

A number of applications have been suggested for carbon nanotubes, particularly in the construction of a variety of mechanical devices. In addition, scientists have found ways to open and fill these nanotubes with a variety of materials, ranging from metals to biological molecules. These nanotubes could have a host of applications, especially as catalysts.

Nanotubes may have even more important applications in electrical devices, however. Imagine, for example, a thin, metallic nanowire within an insulating nanotube. A nanowire is a very thin, nearly solid hair-like nanotube. Then imagine bundling a number of nanotubes together to make an electrically conducting cable. This cable would be stronger, more impermeable, and even more electrically conductive than the best copper cables available today.

In fact, the inclusion of a metal-conducting core would probably not be necessary. Research has shown that the nanotube itself can serve as a conductor. In one series of experiments, Smalley's research team oxidized the end of a nanotube, extracting a long chain of naked carbon atoms. The carbon atoms were remnants of carbon pentagons and hexagons extracted from the basic graphene structure of the nanotube. The team found that a single string of carbon atoms could carry nearly a microampere of electric current. A microampere is not a great deal of current, but if a cable were to consist of, say, 1,000 nanotubes, the total current carried by that cable would amount to more than 1,000 A (amperes) per square centimeter. This is significantly higher than the highest current density ever observed.

Production of Buckyballs and Nanotubes

When striking new breakthroughs in chemistry are made, researchers often warn that future research and development may be hampered because of "the difficulty with which such materials can be prepared." Research on buckyballs and nanotubes represents a contrast to that situation. One of the important features of these materials is the *ease* with which they are formed. Researchers have discovered that when carbon is vaporized, it condenses naturally in the form of fullerenes. Some care is required in the process, however, since graphite does not vaporize at temperatures of less than 3,000°C, and the most favorable temperature range for producing buckyballs is about 1,100°–1,300°C.

The technique now used for making buckyballs, therefore, is to bombard a piece of graphite with a laser beam. The beam raises the temperature of the graphite to at least 3,000°C for a few billionths of a second. The material then cools quickly to about 1,000°C, where it is maintained long enough to permit the formation of fullerenes in yields of 30 to 40%.

The use of nickel and/or cobalt catalysts results in an interesting variation in this experiment. The yield of fullerenes greatly decreases to about 1% in this situation, but more than half of those fullerenes are produced in the form of single-walled nanotubes. It is clear from these results that the production of buckyballs and nanotubes, while not quite as simple as generating oxygen in the laboratory, should present no serious technological challenges to chemists in the future.

References

"A Carbon Nanotube Page," <http://www.rdg.ac.uk:80/~scsharip/tubes.htm> (26 September 1997).

Ebbesen, T. W., "Carbon Nanotubes," *Physics Today*, June 1996, 26–32.

Ebbesen, Thomas W. *Carbon Nanotubes: Preparation and Properties*. Boca Raton, FL: CRC Press, 1996.

Iijima, S., "Helical Microtubules of Graphitic Carbon," *Nature*, 7 November 1991, 56–58.

Iijima, S., and T. Ichihashi, "Single-Shell Carbon Nanotubes of 1-nm Diameter," *Nature*, 17 June 1993, 603–05.

Rudoff, R. S., et al., "Single-Crystal Metals Encapsulated in Carbon Nanoparticles," *Science*, 15 January 1993, 346–48.

Saito, R., et al., "Electronic Structure of Graphene Tubules Based on C-60," *Physical Review B*, 15 July 1992, 1804–11.

Smalley, R. E. "From Balls to Tube to Ropes: New Material from Carbon," address to the South Texas Section of the American Institute of Chemical Engineers, 4 January 1996, <http://cnst.rice.edu/aiche96.html> (26 September 1997).

Tans, S. J., et al., "Individual Single-Wall Carbon Nanotubes as Quantum Wires," *Nature*, 3 April 1997, 474–77.

Thess, Andreas, et al., "Crystalline Ropes of Metallic Carbon Nanotubes," *Science*, 26 July 1996, 483–87.

Yakobson, Boris I., and Richard E. Smalley, "Fullerene Nanotubes: C1,000,000 and Beyond," *American Scientist*, July–August 1997, 324–27.

Dendrimers

An important motivation behind a good deal of chemical research is an element of play. Chemists become interested in molecules that have peculiar and interesting shapes (like cubes, airplane propellers, and ladders) or that behave in unusual and unexpected ways (such as blinking on and off when exposed to light). That motivation has certainly been present in much of the research carried out on a fairly new type of macromolecule known as a dendrimer.

Fundamentals of Dendrimer Construction

Reports on these materials first began to appear in the late 1970s and early 1980s. The term *dendrimer*, from the Latin word for "tree," provides a good image of the molecular structure of these materials. In their simplest form, they consist of a single, basic unit with two or more branches, from which additional branches extend. For example, one type of dendrimer can be produced by reacting 1,4-diaminobutane with four equivalents of acrylonitrile ($CH_2=CHCN$). The product of this reaction has four branches extending off the four-carbon base from the original butane molecule. At the end of each branch is a cyano group ($-CN$), which is easily reduced to an amino group ($-C-NH_2$). Each of the four amino groups produced can, in turn, be reacted with two other molecules, such as two other acrylonitrile molecules. In the next stage of the reaction sequence, then, the molecule consists of its core with eight branches (two new branches on each of the original four branches). This process can be repeated as often as one wishes, producing a 16-branch, 32-branch, 64-branch, and so forth, treelike structure.

This description provides only the simplest introduction to the potential for dendrimer construction. For example, one might use a molecule with three functional groups, such as trimethylamine ($N(CH_3)_3$), allowing for the tripling of the number of branches in each step. (See figure 1.1.)

Evolution of Dendrimer Research

Early pioneers in research on dendrimers have pointed out that part of their work was inspired by the pure fascination of manipulating the molecules of which dendrimers are made. For example, it was of some interest to discover whether there was an upper limit to the size of such structures. To date, research has indicated that there is no such limit. Dendrimers have been constructed with diameters greater than 10 nm and molecular weights of more than 1 million daltons. A dalton is the unit of measure for atomic weights. It is equal to 1/12 the mass of a carbon-12 atom.

In addition, chemists were curious as to the various shapes into which they could make dendrimers. The most obvious shape, of course, is a sphere that grows in diameter as more and more branches are added. But chemists have also found that they can add new branches to produce

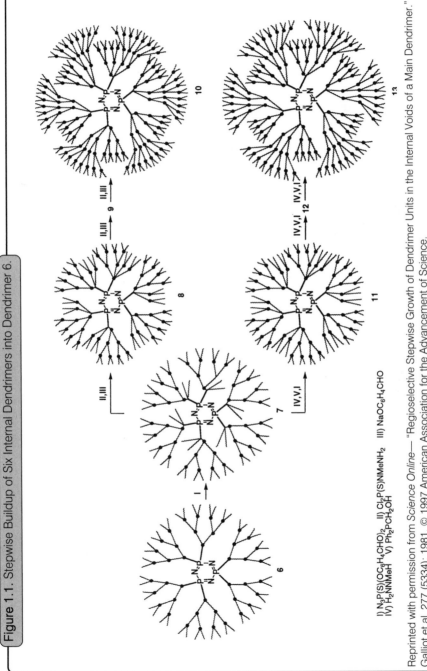

Figure 1.1. Stepwise Buildup of Six Internal Dendrimers into Dendrimer 6.

I) N₃P(S)(OC₆H₄CHO)₁₂ II) Cl₂P(S)NMeNH₂ III) NaOC₆H₄CHO
IV) H₂NNMeH V) Ph₂PCH₂OH

Reprinted with permission from *Science Online*—"Regioselective Stepwise Growth of Dendrimer Units in the Internal Voids of a Main Dendrimer." Galliot et al. 277 (5334): 1981. © 1997 American Association for the Advancement of Science.

boxes, long strings, and hollow, basketlike objects. So far, dendrimers of more than 50 distinctive shapes have been reported.

These shapes often suggest an interesting variety of functions for dendrimer-based materials. For example, in 1994, E. W. Meijer at the Eindhoven Institute of Technology and his colleagues reported on the construction of a hollow dendrimer molecule into which they were able to place a variety of other kinds of molecules. They then discovered ways to "open" the dendrimer molecule to allow the "captured" molecules to pass in and out of the dendrimer. With further research, they were able to control the size and shape of the openings in the dendrimer molecule so that captured molecules could be inserted or released selectively.

In fact, the one term that repeatedly appears in descriptions of dendrimer research is *control*. Although the chemical reactions by which such molecules are produced are usually very complex, it is ultimately possible to precisely control the kinds, numbers, and locations of branches that are added to the core of a dendrimer. Thus, it is theoretically possible to design and manufacture virtually any kind of dendrimer-based product imaginable. In this respect, dendrimers are dramatically different from polymers and most other synthetic macromolecules, which can be made large and complex, but whose precise structures are typically beyond the control of a chemist.

Practical Applications

Over the past two decades, interest in dendrimers has mushroomed. The number of papers has increased from a handful per year in the 1980s to the hundreds per year at the end of the 1990s. One important factor in this growth is that chemists realized the enormous number of potential applications for dendrimers. For example, dendritic molecules with the shape of a box, a hollow cylinder, or a basket might serve as enzymes or tiny containers in which chemical reactions can be carried out. They might also serve as carriers in which drugs are delivered to various parts of a body. As listed in a 1996 report in *Chemical & Engineering News,* other possible applications for dendritic materials include: "magnetic resonance imaging agents, immuno-diagnostics, . . . chemical sensors, information-processing materials, high-performance polymers, adhesives and coatings, separation media, and molecular antennae for absorbing light energy and funneling [it]to a central core"(Dagani).

Recent Research

Research on dendrimers has become so extensive that an adequate review of the field is impossible in a book of this size. A hint of the kind of discoveries being made can be found in research reported in 1997 by a team at the Laboratorie de Chimie de Coordination of CNRS under the direction of Jean-Pierre Majoral. The purpose of this research was to find ways of inserting reactive dendrimer groups within the core of a larger ("main") dendrimer.

The construction of the main dendrimer was accomplished by reacting hexaaldehyde $(CH_3(CH_2)_4CHO)$ with six equivalents of monomethylhydrazine (HN_2NNHCH_3). The product of that reaction was then reacted with six equivalents of a phosphane (Ph_2PCH_2OH) and an azide $(N_3P(S)(OC_6H_4CHO)_2)$ to form a dendrimer with six highly reactive P=N–P(S) fragments within its core. (The abbreviation Ph is sometimes used for the phenyl group, $-C_6H_5$.) This sequence of reactions was then repeated to produce more and more complex forms of the original structure. At this point, the dendrimeric structure was not unlike that of many other "starburst" dendrimers.

In the next step, researchers "sealed off" the outer layer of the dendrimer to prevent further reactions, then attempted to add additional reactive groups *within* the dendrimer sphere. They did so by adding one of two reactive compounds to the dendrimer, anticipating that such compounds would bond with the reactive groups originally installed in the interior of the molecule. The first compound they tried was the azide $N_3P(S)(OC_6H_4CHO)_2$, which produced new branches at the reactive P=N–P segments within the main dendrimer. These branches were then treated with $H_2NN(CH_3)P(S)Cl_2$ and $NaOC_6H_4CHO$ to form a second generation of the main dendrimer, with six branches within its central cavity.

The additive procedure was then repeated with a different combination of reagents: $N_3P(S)(OC_6H_4CHO)_2$, Ph_2PCH_2OH, and $N_3P(S)(OC_6H_4CHO)_2$, which produced a similar result. The researchers concluded their report by stating: "This work demonstrates that functionalization of the internal cavities of various dendrimers can be done through a modification of the skeleton. Various functional groups can be selectively introduced, such as aminophospite, aldehyde, hydrazone, and dichlorophosphane sulfide. Therefore, all the chemistry reported on the surface of dendrimers can now be done in the cavities" (Galliot 1997, p. 1983).

References

Bates, Karl Leif, "A Molecule Made to Order," *The Detroit News*, 27 November 1995; <http://www.detnews.com/menu/stories/26230.htm>.

Dagani, Ron, "Chemists Explore Potential of Dendritic Macromolecules as Functional Materials," *Chemical & Engineering News*, 3 June 1996, <http://pubs.acs.org/hotartcl/cenear/960603/dend.html>.

"Dendrimers," <http://www.melness.demon.co.uk/1996/dendrim.htm> (22 December 1997).

"Dendrimers on the Web," <http://dendrimers.cas.usf.edu/links.html> (22 December 1997).

Galliot, Christophe, "Regioselective Stepwise Growth of Dendrimer Units in the Internal Voids of a Main Dendrimer," *Science*, 26 September 1997, 1981–84.

Tomalia, Donald A., and Roseita Esfand, "Dendrons, Dendrimers and Dendrigrafts," *Chemistry & Industry*, 2 June 1997, <http://ci.mond.org/9711/dend.html>.

Wu, Corinna, "Shaping Synthetic Metals," *Science News*, 21 June 1997, 384–85.

Dihydrogen Bonding

Chemists have traditionally recognized a small number of chemical bonds. Ionic bonds, for example, form between positively and negatively charged ions. Covalent bonds consist of shared pairs of electrons. Hydrogen bonds form when a partially positively charged hydrogen atom bonded to a nitrogen or oxygen atom is attracted to a center of negative charge, such as an oxygen or nitrogen atom on another molecule (intermolecular bonding) or within the same molecule (intramolecular bonding).

Hydrogen bonding is of special interest to chemists partly because of its role in the biochemical molecules that occur within living organisms. A hydrogen bond is just strong enough to allow a molecule to retain its shape, but just weak enough to break when that molecule becomes involved in some chemical reaction. The parallel strands in a DNA molecule, for example, are held together by hydrogen bonds. These bonds allow the DNA molecule to maintain a stable conformation, while allowing it, at specific times, to divide during the processes of replication and translation.

In 1996, a team of researchers headed by Dr. Robert H. Crabtree at Yale University discovered a new kind of chemical bond, a bond they called a *dihydrogen bond*. This discovery came about somewhat unexpectedly as the Crabtree team was studying iridium hydride complexes. Some of the structures they encountered had shapes significantly different from what they had predicted. In particular, some of the bond lengths they observed were much shorter than they expected.

The Aminoboranes

In order to pursue this finding, Crabtree's team searched the Cambridge Crystallographic Database for compounds with short hydrogen-hydrogen bond lengths. The class of compounds in which they became particularly interested were the aminoboranes, compounds with a characteristic B–H $\cdots$ H–N–H structure. A striking property of these compounds was that their melting points were dramatically higher than those predicted based on their structures. For example, the parent compound of that family, aminoborane (H_3BNH_3) has a melting point of 104°C. By comparison, compounds with roughly similar molecular masses, such as ethane (C_2H_6) and monofluoromethane (CH_3F) have dramatically lower melting points (–181°C and –141°C, respectively).

The presence of such high melting points in the aminoboranes suggested that some kind of intermolecular bonding was significantly increasing the amount of energy needed to break the molecules apart to allow the solid compound to become liquid. Hence, the researchers' next task was to determine the nature of this intermolecular bonding.

In his report, Crabtree noted that the problem was resolved through "the close cooperation among synthetic, physical, crystallographic, and theoretical approaches" (p. 348). Experimental observations on the iridium hydride complexes led to theoretical speculation as to the reason for the observed anomalies, which, in turn, led to research on a possible new form of bonding.

Results

In an early phase of their research, the Crabtree team modeled and then experimentally studied the interaction between two aminoborane molecules in the gas phase. They found that a relatively strong force of attraction between any two molecules indicated an H · · · H bond strength of 6.1 kcal/mol. They attributed this bond strength to the formation of a new kind of bond, a *dihydrogen bond*. That bond was hypothesized as being formed between a weakly positive hydrogen atom, as occurs in O–H or N–H, and a weakly negative hydrogen atom, as occurs in boron hydride. These two forms of hydrogen atoms are sometimes referred to as *protonic hydrogen* and *hydridic hydrogen,* respectively. Notice that the major difference between *dihydrogen* bonding and traditional *hydrogen* bonding is that the former occurs between two hydrogen atoms, while the latter involves a hydrogen atom and an electron donor, such as an oxygen or nitrogen atom.

Crabtree and his associates also measured the bond angles within the aminoborane molecule. They found that the bond joining boron to hydrogen to hydrogen to nitrogen was strongly bent. This angled bond differs significantly from a traditional hydrogen bond, which is essentially linear. The reason for the angle in the B–H · · · H–N–H bond appears to be as follows:

The H–N bond in the amino portion of the bond (H–N–H) is strongly polar because of the relatively large difference in electronegativities between the two elements (2.1 vs. 3.0). The hydridic hydrogen attached to boron, then, is relatively strongly attracted to the H–N bond and the bond angle approaches a straight line (160°–180°). In contrast, the H–B bond is only slightly polar because of the very small difference between electronegativities of boron and hydrogen (2.0 vs. 2.1). Thus, the protonic hydrogen of the H–N bond is attracted almost equally to both ends of the H–B bond, producing a sharp bend in the B–H–H region of the B–H · · · H–N–H bond.

Transition Metal Dihydrogen Bonding

Ultimately, the Crabtree team returned to the problem with which this research began: the structure of iridium hydride complexes. Could dihydrogen bonding be used to explain the unexpected structures that had set them off on a search of the Cambridge Crystallographic Database?

One of the experiments designed to answer that question involved the synthesis of a compound in which the potential for dihydrogen bonding existed [$IrH_3(PPh_3)_2$(2-aminopyridine]. That compound was then studied to find out whether H $\cdots$ H attraction of significant magnitude could be detected. The measured bond energy was, in fact, of the order predicted (about 5.0 kcal/mol for the Ir–H $\cdots$ H–N–H bond).

Crabtree's team also explored the possibility of intermolecular dihydrogen bonding. They studied a metal hydride, $ReH_5(PPh_3)_3$, that normally crystallizes poorly from solution and yields small crystals. (Ph is the abbreviation sometimes used for the phenyl group, $-C_6H_5$). When dissolved in indole (2,3–benzopyrrole), however, the metal hydride crystallized as very large, yellow crystals. The team identified the crystals as a complex formed between $ReH_5(PPh_3)_3$ and indole as a result of dihydrogen bonding. As in the aminoborane case, transition metal hydrides appear to have strongly bent bonds in which the X–H $\cdots$ (HM) bond is nearly linear, but the (XH) $\cdots$ H–M bond is strongly bent. (The "M" in this expression represents the transitional metal.)

References

"An Accidental Discovery," *Chemistry and Industry News*, 15 August 1996; <http://ci.mond.org/9615/961504.html>.

Crabtree, Robert H., et al., "A New Intermolecular Interaction: Unconventional Hydrogen Bonds with Element-Hydride Bonds as Proton Acceptor," *Accounts of Chemical Research*, July 1996, 348–54.

Studies of Cubane

In 1964, Philip E. Eaton and Thomas W. Cole, Jr., at the University of Chicago synthesized the compound cubane. As its name suggests, a molecule of cubane consists of eight carbon atoms joined to each other in the shape of a cube. A hydrogen atom is joined to each of the eight carbon atoms, giving cubane the molecular formula C_8H_8.

Cubane is of interest to chemists for a number of reasons. First, it is one of a group of molecules with shapes that mirror those of everyday objects. Other molecules of this kind are propellane, with the shape of a propeller; ladderane, with the shape of a ladder; bowtiediene (or spiropentadiene), with the shape of a bow tie; prismane, with the shape of a prism; and buckminsterfullerene, the allotrope of carbon consisting of 60 carbon atoms arranged in a soccer-ball-like structure. Organic chemists sometimes try to synthesize such structures simply "for the fun of it."

But cubane interests chemists for other reasons, too. For one thing, the carbon-carbon bond angles in cubane are 90°. Every chemistry student "knows" that a carbon atom prefers to form bonds that make an angle of 109.5° with each other, forming the familiar tetrahedral structure on which most organic compounds are formed. Getting carbon to form 90° bonds with each other is a difficult task because of the strain involved. Some chemists had wondered whether it could ever be accomplished in practice. The Eaton/Cole experiment showed that it could.

Properties of Cubane
Cubane is a white, crystalline solid with a melting point of about 132°C. It is remarkably stable at room temperature, with the exception that it has a high vapor pressure and rate of sublimation. The compound is "remarkably stable" because the 90° bond angles represent a high degree of stored energy in carbon-carbon bonds. One might expect that energy to be released relatively easily and, perhaps, even explosively. Instead, the compound appears to be entirely stable at room temperatures.

Cubane has a rhombohedral crystal structure at room temperature. The unit cell of the molecule looks as if one has pushed on one vertex of the cube, distorting its shape to a rhombohedron.

Cubane is a member of a class of solids known as molecular solids. Molecular solids consist of molecules in which the forces between atoms *within* a molecule is much greater than the forces *between* molecules. In general, such molecules tend to pass through a plastic phase just below their melting points. In the plastic phase, the solid slowly deforms and looses its structure without actually melting at some distinct temperature. Their crystal structure in the plastic phase is almost universally a face-centered cube.

The question has been whether cubane behaves like other molecular solids and changes from a rhombohedral structure to a cubic structure in its plastic phase. This question has been difficult to answer because cubane tends to vaporize as it is heated. Thus, no data on the structure of cubane when it is just below its melting point has been available.

In 1997, researchers at the National Institute of Standards and Technology, the University of Chicago, and the University of Maryland found a way to solve this problem. They mixed a small amount of powdered carbon with cubane before heating the compound. They found the added carbon reduced the rate of sublimation enough to allow analysis of cubane's crystal structure in its plastic phase.

To their great surprise, cubane was found *not* to assume a face-centered cubic structure near its melting point. Instead, it retained its rhombohedral structure. Researchers concluded that cubane is "a highly unusual molecular solid" whose further study "could result in new materials with novel properties" (Yildirim 1997, p. 4941).

Practical Applications

Over the past three decades, a number of practical applications for cubane have been suggested. Perhaps the most common has been its potential as an explosive. Although cubane itself is relatively stable, large amounts of chemical energy are stored in its carbon-carbon bonds. Chemists have discussed a variety of ways in which that energy might be increased and then made available "upon demand."

One series of experiments has been designed to produce a derivative of cubane known as octanitrocubane. (See figure 1.2.) Octanitrocubane would retain the core eight-carbon cube in cubane, but would contain eight nitro ($-NO_2$) groups in place of the eight hydrogen atoms in cubane. Nitro groups tend to increase the instability and explosive potential of a compound as, for example, in trinitrotoluene (TNT). Some experts predict that octanitrocubane might be at least twice as powerful as TNT and perhaps even more powerful than HMX (cyclotetramethylene tetranitramine), the most powerful chemical explosive invented since World War II. Octanitrocubane would have the added advantage of being relatively safe and easy to pack, store, and deliver in a warhead.

Some less destructive uses of cubane have also been suggested. Research has shown that cubane may be effective in the treatment of HIV-related disorders, bone marrow cancer, Parkinson's disease, and other medical problems. At this point, however, research on these applications of cubane is not progressing very rapidly.

References

Browne, Malcolm W., "Chemical Cube Packs Power into Beauty," *New York Times*, 15 July 1997, C1+.

Wu, C., "Structure Puts Cubane in a Slanted Box," *Science News*, 12 July 1997, 22.

Yildirim, T., et al., "Unusual Structure, Phase Transition, and Dynamics of Solid Cubane," *Physical Review Letters*, 30 June 1997, 4938–41.

A New Kind of Polymer

It is somewhat remarkable to note that polymers have been an important commercial product for less than about a half century. People who grew up before World War II had virtually no contact with these materials. Names such as polyethylene, polypropylene, polyester, and polyvinylchloride (PVC) had essentially no meaning to people of that era. Yet today, polymers have a nearly endless number of applications in industry and everyday life. They have been studied and produced in a

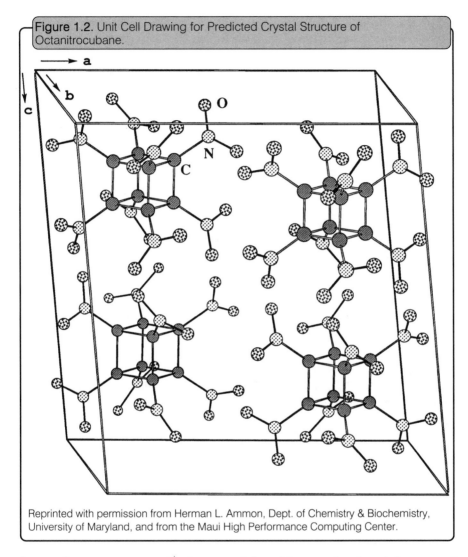

Figure 1.2. Unit Cell Drawing for Predicted Crystal Structure of Octanitrocubane.

Reprinted with permission from Herman L. Ammon, Dept. of Chemistry & Biochemistry, University of Maryland, and from the Maui High Performance Computing Center.

host of new ways to obtain materials with exactly the right sets of properties for many different uses.

For all the variety that exists among polymeric materials, one common chemical characteristic is true of all of them. All synthetic polymers consist of combinations of one or more basic units (monomers) joined to each other by means of covalent bonds. For example, polyethylene can be represented by the general formula –[E–E–E–E–E–E]– where the grouping E–E–E–E represents ethylene molecules joined to each other by covalent bonds. In some cases, two different monomers can be joined to each other, but the bonding in such cases is still covalent bonding.

Hydrogen-Bonded Polymers

In 1997, a research team at the Eindhoven University of Technology in the Netherlands described the formation of a new kind of polymer in which monomeric units are held together by hydrogen bonds. The notion of using hydrogen bonding in polymers would appear to be nonsensical at first glance. Hydrogen bonds tend to be weak bonds that break rather easily when they are warmed or exposed to a variety of chemical environments, such as solutions of different pH levels.

Yet, researchers have explored the possibility of strengthening such bonds by using them in pairs and triplets. Until the Eindhoven experiments, however, such attempts had been largely unsuccessful. Polymers produced with single, double, or triple bonds tend to decompose too easily to have any practical values.

In 1995, the Eindhoven team accidentally discovered that a compound known as 2-ureido-4-pyrimidone has a strong tendency to form dimeric (two-part) structures that are held together by four hydrogen bonds. The hydrogen bonds form between hydrogens of amino groups on one 2-ureido-4-pyrimidone molecule and the amino nitrogens and carbonyl oxygens on a second molecule of the same kind. When 2-ureido-4-pyrimidone units are attached to the ends of other molecules, a polymer can be synthesized that has many of the properties of a covalently bonded structure. The new kind of polymer is known as a supramolecular polymer because it consists of many individual molecules joined to each other by hydrogen bonds rather than covalent bonds.

Properties of Supramolecular Polymers

The properties of supramolecular polymers are very much a function of the conditions in which they exist. For example, at low temperature and in the presence of nonpolar solvents, the hydrogen bonds are not disturbed and the materials are stable and viscous. These properties are similar to those observed for covalently bonded polymers.

Those properties change, however, in the presence of conditions that tend to disrupt hydrogen bonding, such as elevated temperatures and the presence of polar molecules. In such conditions, the hydrogen bonds holding supramolecular polymers together tend to break, and the material degrades to its monomeric state.

The reversible nature of supramolecular polymers represents a new property for polymeric materials. With conventional polymers, the polymer either exists or it doesn't. Covalent bonds either hold the monomeric units in place, or they break and the polymer is destroyed. With supramolecular polymers, the polymer can be created, destroyed, restored, and destroyed repeatedly, simply by altering the conditions in which it is kept.

Practical Applications

The Eindhoven researchers have *not* suggested that this new material is a desirable substitute for most of the existing polymeric materials available commercially. Such a large variety of polymers exist, they say, that adding a polymer with weaker bonding does not add to the supply of products currently available.

They do suggest, however, that supramolecular polymers might find useful applications in the construction of three-dimensional materials. When covalent polymerization is used in the production of such materials, many active groups in the monomeric units remain unreacted. They become trapped within the mass of the polymer before they have an opportunity to react with some other group.

In contrast, the bonds in hydrogen-bonded polymers are constantly forming and re-forming during the polymerization process, providing many more opportunities for complete bonding among all reactive groups within the structure. This property of supramolecular polymers suggests some commercial applications for the material. For example, one use suggested by the Eindhoven group is in coatings and hot melts. A hot melt is an adhesive that is applied by first heating a plastic and then applying it to a surface. Once on the surface, the plastic cools and solidifies quickly. The ability of a supramolecular polymer to change its physical properties quickly under differing temperatures would appear to make it ideal for such an application.

References

Dagani, Ron, "Supramolecular Polymers," *Chemical & Engineering News*, 1 December 1997, 4.

Sijbesma, Rint P., et al., "Reversible Polymers Formed from Self-Complementary Monomers Using Quadruple Hydrogen Bonding," *Science*, 28 November 1997, 1601–04.

Blinking Molecules

The behavior of fluorescent matter in bulk is relatively well understood. When such matter is irradiated with some form of energy, it gives off light. If one could observe individual molecules of a fluorescent material, one would expect to see the absorption-emission process repeated over and over again in various molecules in the sample. The overall effect would be a continuous fluorescent glow from the sample as a whole.

What would happen if one were able to isolate and observe a single molecule as it underwent the process of energy absorption and emission? A technique known as single-molecule spectroscopy (SMS), developed in the early 1990s, makes it possible to answer that question. The technique provides an entirely new and very promising method for analyzing the process of fluorescence. In 1997, chemists at the University of Minnesota and at the Massachusetts Institute of Technology (MIT), under the direction of Paul F. Barbara, reported on the emission spectra of a single fluorescent molecule that contains many chromophore groups.

Expectations and Reality

The molecule studied in this research was a derivative of the copolymer of poly(p-phenylene vinylene) and poly(p-pyridylene vinylene) with a molecular weight of about 20,000. Each molecule of this polymer contains dozens of chromophore groups, which would be expected to fluoresce when the molecule is excited.

The emission spectrum of such a molecule might be expected to be quite complex since excitation of a single molecule might cause emission at any one or more of the chromophore groups in the molecule. The molecule might be expected to behave like a *group of molecules* because of the presence of so many chromophore groups within it.

In fact, no complex spectrum was observed experimentally. Instead, when a single molecule of the polymer was irradiated, all chromophore groups turned on and off almost simultaneously. The many chromophore groups were all emissive or nonemissive (dark) at the same time.

Explanation

How can this highly coordinated blinking behavior be explained? How can the signal for excitation or quenching transit the whole polymer chain in a matter of picoseconds? Barbara's research team suggests that excitation may produce a reversible chemical change within the polymer chain, such as an electron transfer resulting in the formation of a cation-anion pair. The presence of cations in a chromophore group is known to cause a quenching (deactivation) effect in that group. Creation of the cation-anion pair, then, would be quickly followed by the reverse of that reaction, with the transfer of an electron resulting in the neutralization of the two ions.

The intriguing question that remains, then, is how a chemical change at one site in the polymer communicates its effect along the whole chain at virtually an instant. To date, no answer to that question has been found.

References

Vanden Bout, David A., et al., "Discrete Intensity Jumps and Intramolecular Electronic Energy Transfer in the Spectroscopy of Single Conjugated Polymer Molecules," *Science*, 22 August 1997, 1074–77.

Wilson, Elizabeth, "The Case of the Blinking Polymer," *Chemical & Engineering News*, 1 September 1997, 34.

Breaking Individual Chemical Bonds

An important trend in chemical research today is the attempt to study matter and chemical changes at the simplest possible level, that of atoms and molecules. The behavior of materials at this basic level of organization may not only be very different from the behavior of matter at the bulk level, but it also may provide insights into the structure and interactions of matter that are not available on a larger scale.

An example of this research was reported in 1995 by Tsung-Chung Shen and his colleagues at the University of Illinois at Urbana-Champaign. In this study, Shen's group used a scanning tunneling microscope (STM) to break individual chemical bonds in a semiconductor-like material. The material contained sites at which silicon atoms were bonded to single hydrogen atoms (SiH) and to two hydrogen atoms (SiH_2).

A scanning tunneling microscope has a very sharp tip from which tunneling electrical currents are established as the microscope travels over the surface of a material. The images formed by the variation of the current or the gap between the tip and the surface provides information regarding the composition and shape of the surface.

Methodology and Results

Researchers prepared the surface of a silicon crystal doped with either arsenic or boron and with one layer of hydrogen atoms. The STM was then passed over the surface of this material. The voltage of the STM was varied across a range from about 2 V to about 12 V. In this process, the STM had two different functions. First, electrons released at the tip of the STM provided sufficient energy to break Si-H bonds in the surface of the material. Second, the STM took pictures of the surface across which it had moved and acted upon previously.

STM images showed the presence of discrete points on the silicon surface where individual hydrogen atoms had been removed. (See figure 1.3). The presence of "dangling" silicon bonds, created by the loss of hydrogen atoms, showed up in the STM images as bright spots. Researchers were able to count the number of "dangling" bonds and, thus the number of hydrogen atoms lost, for each voltage and current used in the STM. They found that no bonds were broken at voltages of less than 4 V at low current (10^{-11} A), but that bond dissociation increased rapidly to a maximum between 6 V and 8 V. Above about 7 V, the number of bonds broken remained constant.

Investigators were also able to pinpoint the exact location of each broken bond on the material's surface. As one member of the team observed, "We can actually get down to individual rows of molecules on

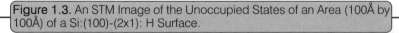

Figure 1.3. An STM Image of the Unoccupied States of an Area (100Å by 100Å) of a Si:(100)-(2x1): H Surface.

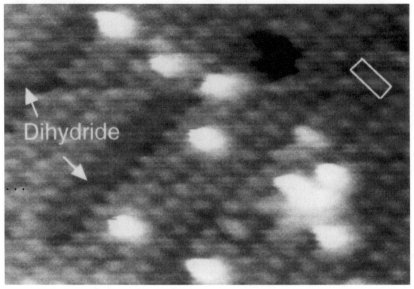

Reprinted with permission from T.C. Shen et al. "Atomic-Scale Desorption through Electronic and Vibrational Excitation Mechanisms." *Science*, Vol 268 (16 June 1995): 1591. ©1995 American Association for the Advancement of Science.

a silicon surface and selectively remove hydrogen along those rows" (Lipkin 1995, p. 391).

Two mechanisms appear to be involved in the breaking of bonds. First, a single energetic electron from the STM may transfer enough energy to electrons in a Si–H bond to cause it to rupture. Second, lower-energy electrons from the STM may provide enough energy to cause individual hydrogen atoms to begin vibrating and, eventually, to break loose from their silicon partners.

Significance

Techniques used to etch surfaces on a larger scale often do not work at the nanometer level. For example, semiconductors of the scale traditionally used can be etched by chemical or photographic techniques. Accomplishing the same objective on the nanometer scale, however, requires the use of new methods. The use of the STM to pluck individual atoms off a surface, described in this report, may provide a technique for etching electronic devices containing components at the molecular level.

References

Lipkin, R., "System Breaks Individual Molecular Bonds," *Science News*, 24 June 1995, 391.

Shen, T. C., et al., "Atomic-Scale Desorption through Electronic and Vibrational Excitation Mechanisms," *Science*, 16 June 1995, 1590–92.

CHEMICAL REACTIONS

Learning more about chemical changes is, of course, one of the primary goals of chemical research. The following selections focus on two general kinds of research. One line of research involves the study of reactions at the atomic scale, which until quite recently was technically very difficult or impossible. The studies reported below describe breakthroughs in the analyses of chemical changes at their very simplest levels.

A second line of research involves the study of familiar chemical reactions under somewhat unusual conditions for those reactions. For example, one study focuses on the reaction between oxygen and hydrogen, one of the most thoroughly studied of all chemical reactions. But the reaction is carried out at extreme conditions of temperature and pressure, with surprising results.

Peculiarities in Methane Clathrate Hydrate Formation

The search for new sources of energy is an ongoing focus of chemical research. In 1996, scientists at the U.S. Geological Survey (USGS) in Menlo Park, California, and the Lawrence Livermore National Laboratory (LLNL) in Livermore, California, made an unexpected and interesting discovery. They found that ice could be superheated to temperatures of nearly 16°C for periods of up to eight hours.

The research was initiated in an attempt to study the physical properties of methane clathrate hydrate. Clathrates are compounds in which one chemical species is enclosed within the cagelike structure of a second chemical species. In this case, methane clathrate hydrate consists of a network of water molecules bonded to each other by means of hydrogen bonds in which molecules of methane gas are trapped. One form of the clathrate contains 46 water molecules, with 8 cavities. Thus, the theoretical formula for this compound is $CH_4 \cdot 5.75\ H_2O$. Note, however, that the clathrates are not stoichiometric compounds, and their actual molecular compositions tend to vary from the theoretical structure somewhat.

The methane clathrates are of some practical interest to scientists because some experts believe that huge amounts of natural gas (primarily methane) are trapped within these compounds in the Earth's surface. Indeed, some authorities have called the methane clathrates "the largest untapped reservoir of natural gas on Earth" (Stern 1996, p. 1843). These clathrates are known to occur in sediments beneath permafrost regions, within the ocean bottoms, and along the continental shelves and continental margins. In addition, methane clathrates are thought to occur on the moons of the outer planets of Saturn, Jupiter, and Uranus.

The Research

The original purpose of the USGS-LLNL research was to determine certain physical properties of the methane clathrate hydrates, including the way the substances respond to stresses and strains placed on them. The first step in the experiment was to produce methane clathrate hydrate by heating together granular ice at about 250 K with methane gas under a pressure of about 35 Mpa (35 megapascal, or 35 million pascal). One would expect the clathrate to form as a solid below the freezing point of water and then, as the temperature increased, to melt and decompose. The researchers detected the formation of methane clathrate hydrate with the empirical formula of $CH_4 \cdot 6.1H_2O$. The process they expected to observe at the melting point of water was approximately $CH_4 6.1 \, H_2O$ (solid, then liquid)$\rightarrow 6 \, CH_4$ (g) $+ 6 \, H_2O$ (l).

The anomaly that was observed was that solid clathrate remained even when the temperature rose more than 15°C above the melting point of water. The researchers' first thought was that they had made an error in the experiment, either in reading their equipment or in carrying out the experiment. To test this hypothesis, they repeated the experiment using neon gas rather than methane. Neon does not form clathrates, so its substitution for methane should indicate whether the equipment itself was working properly. As it turned out, it was. Ice used in combination with neon gas melted completely within 30 minutes of reaching its melting point.

Explanation

The researchers considered a number of explanations for the results they observed with methane clathrate hydrate. So far, however, they do not consider any of their hypotheses adequate to explain their findings. Indeed, the results must be considered tentative until they are confirmed by further research.

References

"Ice Resists Melting in Warm Conditions," *Science News*, 19 October 1996, 252.
Stern, Laura A., Stephen H. Kirby, and William B. Durham, "Peculiarities of Methane Clathrate Hydrate Formation and Solid-State Deformation, Including Possible Superheating of Water Ice," *Science*, 27 September 1996, 1843–48.

Observing Individual Chemical Reactions

Most of the knowledge that chemists have of chemical reactions is based on the study of very large numbers of particles. Even a microliter of solution, for example, contains millions of atoms, molecules, ions, or other species. It would seem unlikely that a single molecule or a single chemical reaction could be observed with current methods of analysis.

Reflecting that empirical reality, chemical theories deal with statistical analyses of large numbers of particles. In any particle reaction, for example, it might be possible to state the average time required for a reaction to occur among many millions of particles. But in any given sample, that reaction may occur more rapidly in some cases and less rapidly in others. In general, such reactions are likely to follow a Poisson distribution that can be described by a bell-shaped curve with a maximum at some given value.

Technology now makes it possible to study individual chemical reactions under certain specialized conditions. In the gaseous phase, for example, reactants can be supplied to a reaction chamber under very carefully controlled conditions, allowing the study of individual reactions. In the solid state, reactants are essentially immobile, so they can be studied by certain kinds of microscopy and spectroscopy.

Solution Reactions
The vast majority of chemical reactions take place in solutions. In such cases, reactants are in rapid motion, colliding with each other and the walls of their container at a pace that makes their behavior difficult to study.

In 1995, Maryanne Collinson and R. Mark Wightman reported on a method by which individual chemical reactions in solution can be observed. The key to their procedure was the use of a very small reaction cell that contains two microelectrodes. The total surface area of the electrodes was 9×10^{-7} cm^2.

The researchers introduced into the cell a solution of 9,10-diphenylanthracene (DPA) dissolved in acetonitrile. They then passed a small electric current between the two electrodes. The current changed sign at regular intervals, first generating positive ions and then generating negative ions.

Assume that during the first electrical pulse, positive ions of DPA are formed:

$$DPA \rightarrow DPA^+ + e^-$$

Once formed, these ions diffuse away from the electrode. Now assume that a few microseconds later, the current changes sign and negative ions are formed:

$$DPA \rightarrow DPA^- - e^-$$

As these ions migrate away from the electrode, they eventually come into contact with the cations formed during the previous electrical pulse. When the two ions combine, they reunite to form the neutral parent:

$$DPA^+ + DPA^- \rightarrow 2\ DPA$$

During this reaction, however, the DPA formed is in an excited state that emits a single photon as it returns to its ground state. The emission of this photon can be used as an indicator that the chemical reaction has occurred.

Experimental Technique

The secret of success for this experiment was to observe the status of all DPA species during the very short time that they existed as ions, which would eventually combine to form a neutral molecule. The researchers achieved this objective by constructing a "time-space" cell of very small dimensions in which the reaction could occur. The space dimension of the cell was the surface area of the electrodes. The time dimension of the cell was equivalent to the distance the ions could travel during the time between electric pulses.

The time during which cations were allowed to form in the reaction was 500 μs (microseconds or 10^{-6} seconds). During this time, cations formed and migrated away from the electrode surface. The time permitted for anion formation, however, was only 50 μs. This condition meant that anions could travel for only 50 μs before they encountered a cation formed in the first step. The effective size of this cell was about 20 femtoliters.

Results

The results obtained in this experiment depended on the time scale on which they were measured. By taking observations once every second, the researchers found an essentially continuous release of luminescence from the DPA. By reducing the time limits over which light pulses were measured, however, a more detailed view of the reaction was possible. Taking measurements every microsecond, for example, revealed a series of pulses taking place every 0.5 μs. This series of pulses could be construed as dozens or hundreds of individual pulses in close proximity to each other.

At even smaller time intervals, pulses could be observed in even more discrete packages. When light pulses were measured every 100 ns, for example, groups of closely packed lines at 0 μs, 600 μs, 1200 μs, etc., could be observed. Finally, at the finest level of resolution, individual light pulses could be observed. This level of resolution occurred when pulses were measured every 5 ns. At this level of resolution, discrete pulses could be observed, for example, at 33.8 μs, 34.2 μs, 34.3 μs, 34.5 μs, and so forth. Each of these pulses represented the release of a single photon of light, which, in turn, represented a single reaction between ions of DPA to form the neutral molecule.

Significance
One important finding in this research was that the characteristics of the reaction studied closely matched those predicted by theory. Although chemists have long had a good deal of confidence in their theories of chemical kinetics, it was exciting to have these theories confirmed by experimental observation.

In addition, the technique developed by Collinson and Wightman appears to have application to many other kinds of chemical reactions. The analysis they used here is not only effective, but also very simple to carry out.

References
Collinson, Maryanne M., and R. Mark Wightman, "Observations of Individual Chemical Reactions in Solution," *Science*, 30 June 1995, 1883–84.

Lipkin, R., "Observing Individual Molecular Reactions," *Science News*, 15 July 1995, 38.

Reaction of Hydrogen and Oxygen at High Pressure
Every high school and college chemistry student is familiar with the reaction in which hydrogen and oxygen react to form water:

$$2\,H_2 + O_2 \rightarrow 2\,H_2O - \Delta H$$

Chemists have been studying this reaction for more than 200 years. Still, some details of the change are not totally understood. The energy term in the equation indicates that large amounts of energy are released when the two elements combine.

At least, that's what happens under ordinary conditions of temperature and pressure. But what happens when pressure on the reactants is increased?

Most students and professional chemists would be likely to give the same answer to that question. In general, an increase in pressure tends to increase the number of molecular collisions and, hence, the rate at which the reaction occurs. The combination of hydrogen and oxygen at very high pressures, then, might be expected to result in a much more rapid formation of water than would occur at atmospheric pressure.

In 1995, two scientists at the Université Paris, Paul Loubeyre and René LeToullec, reported strikingly different results from those previously observed or expected for the reaction between hydrogen and oxygen at high pressures. At pressures of 76,000 atm, the 2 elements did *not* react explosively at all, nor did they form water as a final product.

Methodology

Researchers exposed a mixture of oxygen and hydrogen gases at very high pressure in a diamond anvil cell. A diamond anvil cell is a small device, easily held in the hand, in which pressures of more than 400,000 atmospheres can be produced. In the Paris experiments, the cell had a diameter of 50 μm and an initial thickness of about 15 μm. Pressure on the gases was increased gradually, while temperature was left constant at room temperature.

A reminder of the sensitivity of this reaction occurred during one loading when the two gases combined explosively, causing major damage to the researchers' equipment. The cause of the explosion, they suggested, may have been "too fast a pressure increase" (Loubeyre 1995, p. 44) on the gases.

Results

Results obtained in the experiment depended, to some extent, on the original composition of the hydrogen/oxygen mixture. When a 10% by mole oxygen mixture was used, a well-defined phase consisting of pure hydrogen could be identified at about 8,000 atm. With a 90% by mole oxygen mixture, a phase consisting of pure, solid oxygen developed above 8,000 atm.

The intriguing finding in any of these cases was the presence of a second phase, which became more prominent with more equimolar concentrations of hydrogen and oxygen. In such mixtures, researchers observed the presence of two solids and one fluid at the eutectic (lowest melting point) pressure of 7.9 GPa. As pressure on the system was increased, the two solids increased in size until they filled the reaction chamber. Researchers were able to obtain high-quality photomicrographs of the crystals formed by this process in the diamond anvil cell.

Raman measurements with an argon laser provided identification of the two solids. One was pure hydrogen, while the other appeared to be an unknown substance that showed emission properties of both oxygen and hydrogen. Researchers concluded that the second solid consisted of an alloy-like substance with the composition $(O_2)_3(H_2)_4$. That is, the solid consisted of a 14-atom structure in which oxygen and hydrogen retained their diatomic structures.

Over the month-long period of the experiment, there was virtually no evidence of the formation of water. Even when energy was supplied to the reaction chamber in the form of a laser beam, a halogen lamp, and a synchrotron X-ray beam, no reaction was initiated. The oxygen-hydrogen "alloy" remained stable at pressures close to 8 GPa for the entire duration of the experiment.

Significance

Loubeyre and LeToullec pointed out two possible applications for their research. First, their results suggest that oxygen and hydrogen can be stored at very high pressures, in a very dense state, for a variety of energy applications, such as in rocket fuel systems or in commercial devices that would be needed in any future hydrogen economy. Second, understanding of the new oxygen-hydrogen alloy could provide clues to the composition of the jovian planets and their satellites. In the core of such bodies, elements and compounds are thought to exist under very high pressures similar to those used in this experiment. The assumption has long been that any oxygen on these planets exists only in the form of water because of these high pressures. It now appears that some may also be present in an elemental form similar to the $(O_2)_3(H_2)_4$ alloy.

References

"Chemistry Can Change Unexpectedly at High Pressures," <http://newton.ex.ac.uk/aip/glimpse.txt/physnews.248.2.html> (21 September 1997).

Hemley, Russell, J., "Turning off the Water," *Nature*, 2 November 1995, 14.

Lipkin, R., "Squeezing H_2 and O_2 Yields New Compound," *Science News*, 4 November 1995, 293.

Loubeyre, Paul, and René LeToullec, "Stability of O_2/H_2 mixtures at high pressure," *Nature*, 2 November 1995, 44–46.

Chemical Reactions under Ambient Conditions

An ongoing challenge for chemical researchers is finding mechanisms by which chemical reactions with commercial importance (or potential commercial importance) can be carried out under moderate conditions of temperature and pressure. For example, the only commercially significant process for the manufacture of ammonia from nitrogen and hydrogen currently available requires temperatures of a few hundred degrees Celsius and pressures of more than 100 atmospheres. Obviously, maintaining these conditions contributes substantially to the cost of ammonia synthesis. The two studies reported below illustrate the kind of progress being made in the effort to carry out important chemical reactions at moderate conditions.

Carbon-Hydrogen Bond Activation at Room Temperatures

Every beginning student of organic chemistry learns about the relative inactivity of members of the alkane family. The strength of the carbon-hydrogen (C–H) bond, they are told, accounts for the fact that alkanes typically react only under extreme conditions of temperature and pressure or in the presence of highly reactive species (such as halogens), radiation, or catalysts. That fact also means that the kinetics of alkane reactions are difficult to study and, on a practical level, present problems for industrial applications of such reactions.

In 1997, chemists at the University of California at Berkeley and at the Lawrence Berkeley National Laboratory reported on an important break-through in the study of carbon-hydrogen bond activation reactions. The success of this research depended on two factors: the use of a complex metal-organic compound that reacts essentially stoichiometrically with alkanes and the use of ultrafast spectroscopic instruments that allowed the study of reaction stages at the femtosecond, picosecond, and nanosecond levels.

The compound used in this research was a metallic complex in which dicarbonyl rhodium ($Rh(CO)_2$) is chelated with tripyrazolyl borate ligands. The structure of that compound can be represented as

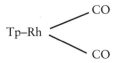

where Tp represents a complex of three 3,5-dimethylpyrazolyl groups attached to a single boron atom.

Stages of the Reaction
The stages that occur in the reaction between this compound and an alkane (represented by R-H) are described below. Each stage was studied spectroscopically with equipment that allowed the identification of the structures formed during the reaction at extraordinarily short time intervals (measured in femtoseconds, picoseconds, and nanoseconds).

1. The original rhodium complex is excited with ultraviolet radiation. In a period of less than 100 femtoseconds, the complex dissociates, with the loss of a carbonyl group:

$$Tp\text{--}Rh\big\langle{}^{CO}_{CO} \rightarrow Tp\text{--}Rh\text{--}CO + CO$$

2. The activated product molecule then adds an alkane molecule at the site formerly occupied by the lost carbonyl group:

$$Tp\text{--}Rh\text{--}CO + RH \rightarrow Tp\text{--}Rh\text{--}CO \cdots R\text{--}H$$

This process occurs within a few picoseconds after the loss of the carbonyl group.

3. The original tripyrazolyl borate ligand breaks apart, leaving the rhodium atom attached to only two of the pyrazolyl groups:

$$Tp \cdots Rh–CO \cdots R–H.$$

This process takes about 200 picoseconds.

4. The dechelation of the ligand apparently provides sufficient energy to cause cleavage of the C-H bond in the alkane, permitting the formation of new bonds between Rh and R and Rh and H.

$$Tp \cdots Rh–CO + R–H \rightarrow Tp \cdots \overset{\displaystyle R}{\underset{\displaystyle H}{Rh–CO}}$$

The bond activation step takes about 230 nanoseconds.

5. Finally, the Tp ligand rechelates to form the original rhodium complex that now contains the products of alkane decomposition rather than the original dicarbonyl form:

$$Tp \cdots \overset{\displaystyle R}{\underset{\displaystyle H}{Rh–CO}} \rightarrow Tp–\overset{\displaystyle R}{\underset{\displaystyle H}{Rh–CO}}$$

Significance

The significance of this research is that chemists now have a very detailed and precise description of one mechanism by which C–H bonds can be broken under the moderate conditions of room temperature. This information may suggest ways by which catalysts can be designed to make possible alkane reactions for practical applications, such as the conversion of petroleum components under room temperature conditions.

References

Brombert, Steven E., et al., "The Mechanism of a C–H Bond Activation Reaction in Room-Temperature Alkane Solution," *Science*, 10 October 1997, 260–63.

Rouhi, Maureen, "Real-World C–H Bond Activation," *Chemical & Engineering News*, 13 October 1997, 4–5.

Cleavage of Dinitrogen at Ambient Conditions

One of the most abundant natural resources on Earth is being utilized by human chemists to only a moderate degree and with relatively great difficulty. Yet, Mother Nature uses that same resource extensively and with comparative ease for some of the most basic chemical reactions that take place in natural processes. That natural resource is elemental nitrogen.

Nitrogen is by far the most abundant gas in the Earth's atmosphere, making up about 78% of the atmosphere by mass. The dinitrogen molecule (N_2) is very stable, with a heat of formation of 944 kJ/mol. The stability of this molecule can be attributed to the triple bond joining the two nitrogen atoms in the molecule.

Certain organisms have evolved mechanisms for cleaving the N–N triple bond, thus making the individual atoms available for reacting with other chemical species. These organisms contain enzymes that catalyze the conversion of elemental dinitrogen to a combined form, such as ammonia (NH_3). This process is known as nitrogen fixation.

Human chemists have been less successful in learning how to cleave the N–N triple bond. One important exception is the Haber-Bosch process for producing ammonia from elemental nitrogen and hydrogen. However, that process occurs only under extreme conditions of temperature and pressure. In contrast, natural nitrogen fixation occurs at ordinary conditions of temperature and pressure. Thus, chemists have long been interested in finding a mechanism by which elemental nitrogen can be utilized in industrial processes under more moderate and, therefore, less expensive conditions.

Dinitrogen Cleavage Using a Molybdenum Complex

In 1995, Catalina E. Laplaza and Christopher C. Cummins at the Massachusetts Institute of Technology (MIT) reported a mechanism by which this goal can be achieved. This mechanism involved the use of a three-coordinate complex of trivalent molybdenum with the formula $Mo(NRAr)_3$, where R represents $C(CO_3)_2CH_3$ and Ar represents 3,5-$C_6H_3(CH_3)_2$. When this compound was recrystallized in a solution of ethyl ether at -35°C in an atmosphere of nitrogen gas, an intense purple color developed. Researchers attributed this color to the formation of an intermediary complex, in which two molecules of $Mo(NRAr)_3$ are linked to each other through a dinitrogen bridge: $Mo(NRAr)_3 \cdots N \equiv N \cdots Mo(NRAr)_3$.

As the ether solution was warmed to 30°C, the purple color began to disappear and was replaced by a gold color. The new compound was identified by spectroscopic analysis as a hexavalent nitrido complex of molybdenum with the formula $Mo(NRAr) \equiv N$. The reaction had resulted, therefore, in the cleavage of the original $N \equiv N$ bond.

Laplaza and Cummins attributed the kinetics of this reaction to the strength of the $Mo \equiv N$ bond in the final product. The very low dissociation enthalpy of this bond, they argue, provides the driving force needed to cleave the dinitrogen molecule.

Significance
The critical feature of this discovery for potential practical applications is that the reaction takes place readily at temperatures ranging from –35°C to +30°C, at pressures of 1 atm, and in relatively short periods of time (less than 1 hour for each of the 2 steps in the overall reaction).

References
Laplaza, Catalina, and Christopher C. Cummins, "Dinitrogen Cleavage by a Three-Coordinate Molybdenum(III) Complex," *Science*, 12 May 1995, 861–62.
Lipkin, R., "Nitrogen: Breaking It up is Hard to Do," *Science News*, 13 May 1995, 294–95.

Femtosecond Observations

The precision with which scientists observe atomic and molecular changes continues to improve. The ultimate goal of such research is to observe such changes as they occur, somewhat as time-lapse photography allows botanists to observe the detailed steps in the blossoming of a flower. In the case of atoms and molecules, however, the "film" would record specific, discrete changes in atoms and molecules as they vibrate, rotate, move from one place to another, and undergo other physical changes.

The rate at which such changes take place is of the order of a few femtoseconds (10^{-15} s). For example, the period of vibration for a molecule is a few femtoseconds. In 1997, a group of French researchers reported on a new technique to follow changes that occur in the femtosecond range. The team was lead by Antoine Rousse at the Laboratorie d'Optique Appliquée, ENSTA—Ecole Polytechnique of CNRS and included researchers from that laboratory, from the X-ray Optics Group of the Institute of Optics and Quantum Electronics at Friedrich Schiller University in Jena, Germany, the Laboratorie de Physique des Solids in Orsay, France, and the Laboratorie pour l'Utilisation des Lasers Intenses at CNRS.

Theory and Experiment
The principle behind the Rousse experiment was to initiate some change in a material and then to observe that change in the shortest possible time period. The equipment used in the experiment included a thin target of cadmium arachidate. Cadmium arachidate is the salt of cadmium metal and an organic acid, arachidic acid, found in peanut oil.

The molecular structure of such a target can be studied by means of X-ray diffraction techniques. In X-ray diffraction, a beam of X-rays is shined on a target. The X-rays are then diffracted by the rows of particles that make up the sample, much like the way a light beam is diffracted by the lines in a diffraction grating. The diffraction pattern depends on a number of factors, such as the number and type of particles and the distances between them within the sample. The diffraction pattern produced by the sample is then recorded by some kind of recording device, such as a camera.

The Rousse team used a laser beam split into two parts in its experiment. One part of the beam was aimed directly at the sample. When the beam struck the sample, it heated up the cadmium arachidate, causing its thermal motion to increase. In such a case, one might expect atoms, ions, and molecules to move more rapidly and for the sample to melt, evaporate, or, at the least, change its crystal structure.

The second part of the laser beam was aimed at a silicon target which, when struck by the beam, gave off X-rays. These X-rays were then also directed at the cadmium arachidate target, thus producing a diffraction pattern.

The ability of this system to measure very brief changes in the target was due to the researchers' ability to alter the time lapse between the two beams. If X-rays struck the target at the exact moment that the laser beam struck, there would be no opportunity to observe any changes in the material. Because the arrival of the X-rays at the target could be delayed until a few femtoseconds after the laser beam hit the target, however, changes that had taken place within that brief moment could be discerned from the changes in the diffraction pattern produced. Further, the amount of time delay would allow researchers to track the progress of such changes by using periods of, for example, x fs, $2x$ fs, $3x$ fs, and so on.

Observations made by the Rousse team allowed them to follow changes in the cadmium arachidate film over periods of no more than a few femtoseconds. They were able to observe increasing disorder in the structure of the film that eventually lead (after a few hundred picoseconds) to the complete evaporation of the film. They concluded from these results that "large structural rearrangements can occur on the femtosecond time scale in condensed matter" (Rischel 1997, p. 492).

References

Jacoby, Mitch, "Femtosecond X-ray Diffraction," *Chemical & Engineering News*, 8 December 1997, 5.

Rischel, Christian, "Femtosecond Time-Resolved X-ray Diffraction from Laser-Heated Organic Films," *Nature*, 4 December 1997, 490–92.

Bubble-Driven Chemical Reactions

Any number of experimental conditions can be employed to bring about a chemical reaction. Heat, pressure, light and other forms of radiation, and even sound are some common examples. In general, any mechanism that produces sufficient energy to break chemical bonds has the potential for initiating a chemical change.

In 1997, scientists at the University of Illinois at Urbana-Champaign, working under the direction of Kenneth S. Suslick, reported on a new approach to this process: the use of hydrodynamic cavitation. Cavitation is the process by which bubbles form, grow, and then collapse in a liquid. The process has been studied for a number of years at least partly because of its practical significance. Engineers have long known that cavitation can be responsible for physical and chemical damage to ship propellers, pumps, and piping systems.

Suslick's research team took advantage of devices that had only recently become commercially available: jet fluidizers capable of driving two jets of fluid towards each other at very high rates of speed. The velocities attained in such devices approach 200 m/s and result in pressure changes as great as 2 kbar.

Chemical Reactions in Jet Fluidizers
Suslick's team explored how the pressures produced in a jet fluidizer might affect a familiar chemical reaction. The reaction they chose is one that has been widely studied in research on *sonochemistry,* the study of chemical reactions induced by sound. The reaction involved the conversion of iodide ion (I^-) to triiodide (I_3^-) in an aqueous solution saturated with carbon tetrachloride. Researchers examined the rate at which this reaction proceeded across a hydrodynamic pressure range between 100 and 1,500 bar.

They found that no chemical change could be detected at pressures less than 150 bar. Above this threshold level, however, the rate at which I_3^- was formed from I^- increased linearly with pressure. They concluded that the characteristics of this reaction as a result of hydrodynamic cavitation were similar to those observed for acoustic cavitation. The study seems to suggest that temperatures produced instantaneously by the collapse of bubbles within the jet fluidizer are sufficient to cause chemical bonds to break.

References
Suslick, Kenneth S., Millan M. Mdleleni, and Jeffrey T. Ries, "Chemistry Induced by Hydrodynamic Cavitation," *Journal of the American Chemical Society*, 1 October 1997, 9303–04.
Wu, C., "Bursting Bubbles Break Chemical Bonds," *Science News*, 11 October 1997, 228.

BIOLOGICAL MOLECULES

Many chemists see chemistry as the science of the twenty-first century. One reason for this opinion is the hope that chemistry will begin to solve many of the fundamental puzzles of other sciences, including biological, geology, and astronomy. Without doubt, a very large amount of research in chemistry over the past decades has focused on developing a better understanding of the molecules of which living organisms are made. The following reports illustrate this line of research.

The Structure of Green Fluorescent Protein

A number of organisms have evolved the ability to produce light of various colors. The best-known example of these organisms may be the firefly. Light production clearly serves some useful function for these organisms, although the exact nature of that function may not always be obvious to humans. In any case, learning *how* an organism generates light can be of value to humans. For example, the mechanism by which fireflies produce light (which is still not entirely understood by scientists) can be used to make tobacco crops produce their own light and shine in the dark.

Transferring the ability to manufacture light from one organism to another is more than an interesting intellectual puzzle, however. Genes that code for light production can have useful applications in research, diagnosis, and other situations. For example, a scientist might want to track the progress of a specific chemical compound as it passes through a series of chemical, physical, or biological changes. One way to do that would be to "tag" the compound with a gene that codes for the production of light. The compound could then be tracked by the light it gives off as it makes its way through a system.

Green Fluorescent Protein (GFP)

One of the most thoroughly studied of all light-emitting compounds is green fluorescent protein (GFP), found in the jellyfish *Aequoria victoria*, which lives off the coast of the Pacific Northwest. GFP is responsible for the greenish glow given off by these jellyfish in their natural habitat. The compound has become the subject of intense study by scientists, at least partly because of its potential use as a chemical and biological marker.

One goal of research on GFP has been to find ways of modifying its structure so as to cause it to emit colors of light other than green. A marker that could be prepared in red-emitting, green-emitting, blue-emitting, and other forms would be enormously useful in tracking a variety of chemical species involved in various chemical and biological processes.

The first step in developing such mutants, however, is to discover the structure of GFP. In 1996, two research teams announced a structure for the GFP molecule. One team was led by S. James Remington, at the University of Oregon, and the other by Fan Yang, at Rice University.

The structure discovered has been compared, somewhat ironically, to the shape of a paint can. (See figures 1.4–1.7.) The compound's 238 amino acid residues are arranged cylindrically, with small caps on the top and bottom of the cylinder. Buried within the middle of the cylinder is the chromophore, situated at an angle of 60° to the central axis of the cylinder. The outside cylinder itself is made up of 11 strands, each containing 9 to 13 amino acid residues. The cap on the top of the cylinder consists of 3 short (5–6 residues) helical segments, and the cap on the bottom consists of 1 such segment.

Light Production in GFP
The key question in this research, of course, is how the chemical structure of the chromophore and surrounding cylinder are related to the production of light by the molecule. A related question might be to ask how

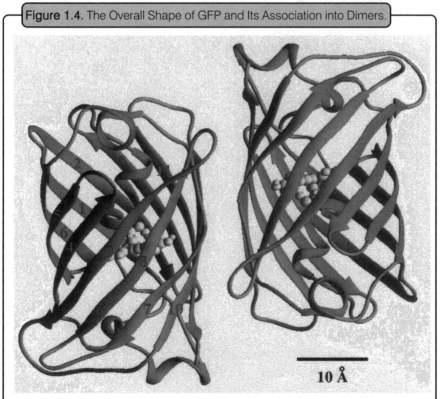

Figure 1.4. The Overall Shape of GFP and Its Association into Dimers.

10 Å

Reprinted with permission from *Nature,* Fan Yang, Larry G. Moss, and George N. Phillips, Jr. "The Molecular Structure of Green Fluorescent Protein." *Nature Biotechnology,* vol. 14, no. 10 (October 1996). © 1997 Macmillan Magazines Limited, and from George Phillips, Jr., Rice University.

Figure 1.5. The Dimer Contact Region.

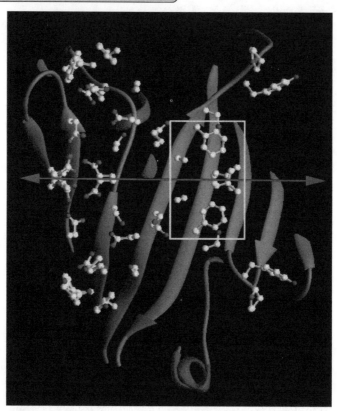

Reprinted with permission from *Nature,* Fan Yang, Larry G. Moss, and George N. Phillips, Jr. "The Molecular Structure of Green Fluorescent Protein." *Nature Biotechnology,* vol. 14, no. 10 (October 1996). © 1997 Macmillan Magazines Limited, and from George Phillips, Jr., Rice University.

those structures can be modified to cause the emission of red, yellow, blue, or some other color (or shade) of light.

Color production is a function not only of the chromophore structure itself, but also of the bonding between the chromophore and cylinder. This principle has led to the manufacture of various mutants of GFP in which one or more amino acid residues is substituted for one or more residues in the natural form of the molecule. These substitutions are designed to increase or decrease the bond strength between chromophore and cylinder. The molecules are then tested to see how the substitutions affect their emission spectra.

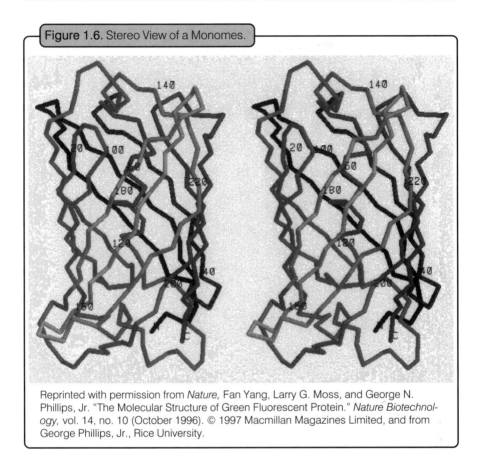

Figure 1.6. Stereo View of a Monomes.

Reprinted with permission from *Nature,* Fan Yang, Larry G. Moss, and George N. Phillips, Jr. "The Molecular Structure of Green Fluorescent Protein." *Nature Biotechnology,* vol. 14, no. 10 (October 1996). © 1997 Macmillan Magazines Limited, and from George Phillips, Jr., Rice University.

As an example, the Remington team successively replaced a threonine residue at position 203 in the cylinder with histidine, tyrosine, and tryptophan. Because the three substituted amino acids are all more polar than the original threonine, it seemed likely that these substitutions could alter the excitation energy of the chromophore.

As it turned out, these changes were successful in modifying both the excitation and emission patterns of the GFP molecule, but only to a relatively modest degree. The most successful mutation produced a slightly more intense green color, but one that was, according to the Remington team, "sufficiently different from those of previous GFP mutants to be readily distinguishable by appropriate filter sets on a fluorescence microscope" (Ormö 1996, p. 1395).

References

Ormö, Mats, et al., "Crystal Structure of the *Aequorea victoria* Green Fluorescent Protein," *Science,* 6 September 1996, 1392–95.

Figure 1.7. Model of the Fluorophore and Its Environment Superposed on the MAD-phased Electron Density Map at 2.2Å Resolution.

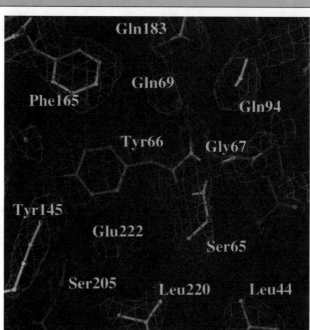

Reprinted with permission from *Nature,* Fan Yang, Larry G. Moss, and George N. Phillips, Jr. "The Molecular Structure of Green Fluorescent Protein." *Nature Biotechnology,* vol. 14, no. 10 (October 1996). © 1997 Macmillan Magazines Limited, and from George Phillips, Jr., Rice University.

Wu, C., "The Shape of More Colors to Come," *Science News,* 5 October 1996, 212.

Yang, Fan, Larry G. Moss, and George N. Phillips, Jr., "The Molecular Structure of Green Fluorescent Protein," *Nature Biotechnology,* October 1996, 1246–52.

The Structure of Some Isoprenoid Enzymes

Chemical equations representing changes that take place in living organisms may seem simple to write and to understand, but those equations seldom give a complete picture of the complex sequence of events that actually take place in the cells of an organism. Ongoing research on the structure of enzymes that catalyze such changes only serves to emphasize this point.

In 1997, the molecular structures of three isoprenoid cyclases were announced. The isoprenoids are a very large and complex class of compounds that contain one or more (and generally, many more) isoprene molecules. Isoprene has the molecular formula $CH_2=C(CH_3)-CH=CH_2$. Some examples of the isoprenoids include the terpenes (which include compounds responsible for the odors of pine sap, basil, bayberry, roses, lemongrass, peppermint, caraway, dill, and camphor; natural rubber; b-carotene; and vitamin A), reproductive hormones, defensive agents, pheromones, membrane constituents, and a host of other important materials.

At least partly because of their widespread occurrence, the isoprenoids have become the focus of intense study by chemists. What, precisely, is the mechanism by which isoprene-related units condense, and how does cyclization (so common in the higher isoprenoids) come about?

Theory
A number of theories have been proposed to answer these questions. According to a currently popular theory, a carbon atom deficient in electrons on one molecule is attracted to an electron-rich region of a carbon-carbon double bond in a second molecule. A new bond forms between the partially positive and partially negative carbon regions, and a chain is extended or closed.

Theories of this kind are only partially satisfactory, however, because they do not explain the physical processes by which molecules are placed in exactly the right configurations for such changes to come about. Nor are they especially helpful in understanding the physical mechanisms by which proton or electron transfers occur. The answers to such questions can be obtained more fully by uncovering the molecular structures of enzymes responsible for such changes in living cells.

Molecular Structure of Squalene Cyclase
An example of this approach can be found in a series of three 1997 reports on the structures of isoprenoid enzymes published in *Science* magazine. One of these reports, on the structure of a squalene cyclase, by researchers at the Institut für Organishe Chemie und Biochemiein Freiburg im Breisgau, Germany, is discussed below. (See figure 1.8.)

Squalene cyclases are enzymes that catalyze the cyclization of an isoprene known as *squalene,* which is found in human sebum (the oily secretion from hair follicles) and in shark-liver oil. Squalene cyclase introduces a twist into a straight-chain portion of a squalene molecule, thus creating a closed ring that did not previously exist there. One step in the cyclization of squalene involves the conversion of the squalene molecule to a reduced form, known as *hopene,* by adding a proton to a carbon-carbon double bond in squalene. The structure discovered for squalene cyclase (S-cyclase) helps explain how this process occurs.

Figure 1.8. Stereoview of Squalene-Hopene Cyclase Chain Fold with Labeled NH$_2$ = and COOH-termini (N and C), Inhibitor Position (L), and Channel Entrance (E).

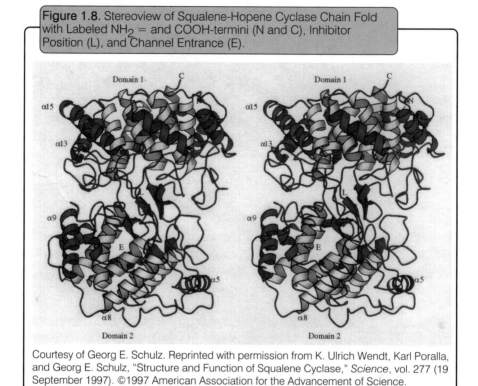

Courtesy of Georg E. Schulz. Reprinted with permission from K. Ulrich Wendt, Karl Poralla, and Georg E. Schulz, "Structure and Function of Squalene Cyclase," *Science*, vol. 277 (19 September 1997). ©1997 American Association for the Advancement of Science.

S-cyclase is a dumbbell-shaped molecule containing two large subunits of 631 amino acid residues each. The protein chain begins at the top of the upper subunit, runs through that unit into the lower unit, and then returns to the upper unit, where it terminates. The active site of the enzyme is located in a large, central cavity located between the two subunits. A narrow channel leads from the central cavity to the exterior of the molecule. A constriction within the channel appears to act as a gate through which a substrate molecule can move into and out of the active site.

The mechanism hypothesized for S-cyclase activity is as follows: A substrate molecule from outside the enzyme moves up the channel, through the gate, and into the active site. The shape of the active site forces the substrate molecule to assume a position in which protonation can occur. That position places the number-three carbon atom in squalene adjacent to a protonated form of an aspartate residue at the top of the central cavity. In that position, the aspartate residue can donate a proton to squalene, converting it to its hopene form. In this form, the molecule leaves the active site and the enzyme, ready for the next step in the chain-elongation and/or cyclization sequence.

References

Lesburg, Charles A., et al., "Crystal Structure of Pentalene Synthase: Mechanistic Insights on Terpenoid Cyclization Reactions in Biology," *Science*, 19 September 1997, 1820–23.

Sacchettini, James C., and C. Dale Poulter, "Creating Isoprenoid Diversity," *Science*, 19 September 1997, 1788–89.

Starks, Courtney, M., et al., "Structural Basis for Cyclic Terpene Biosynthesis by Tobacco 5-epi-aristolochene Synthase," *Science*, 19 September 1997, 1815–19.

Wendt, K. Ulrich, Karl Poralla, and Georg E. Schulz, "Structure and Function of a Squalene Cyclase," *Science*, 19 September 1997, 1811–15.

ATP Synthase as a Nano-motor

The synthesis of adenosine triphosphate (ATP) is one of the most important chemical reactions that takes place within cells. ATP is a major energy-carrying molecule that provides the energy needed to drive many essential biochemical functions. ATP is generated in a series of steps. The first step is the addition of an inorganic phosphate group (P_i) to adenosine monophosphate (AMP), which produces adenosine diphosphate (ADP):

$$AMP + P_i \rightarrow ADP$$

The second step is the addition of a second inorganic phosphate group to the ADP, which produces adenosine triphosphate (ATP):

$$ADP + P_i \rightarrow ATP$$

Because of the biological importance of this reaction, it has been intensively studied for more than three decades. In 1997, the Nobel Prize in Chemistry was awarded to Paul D. Boyer and John E. Walker for their research on the mechanism by which ATP is generated in cells.

Boyer hypothesized in the 1950s and Walker confirmed in the 1990s that the formation of ATP is catalyzed by an enzyme originally known as $F_0F_1ATPase$, which is now known as *ATP synthase*. They found that this enzyme consists of two primary segments, designated as F_1 and F_0. The F_0 segment of the enzyme is attached to a cell membrane, while the F_1 segment is attached to the F_0 segment, but projects outward, away from the membrane.

The F_1 segment of the enzyme consists of three kinds of subunits designated as alpha (a), beta (b), and gamma (g) units. Three alpha and three beta units (a_3b_3) make up a kind of cylinder attached in an immovable position to the membrane. A single gamma unit runs through the center of the a_3b_3 unit. The gamma unit is free to rotate within the a_3b_3

cylinder, driven by hydrogen ions that flow from the membrane into the F_0 unit of the enzyme.

The three beta units within the a_3b_3 cylinder are asymmetrical and have different functions in the synthesis of ATP. As the gamma unit rotates within the cylinder, three different beta units are exposed to the surrounding medium and a single step in the conversion of ADP to ATP takes place within each unit. The 1997 Nobel Prize in Chemistry was awarded to Boyer for his hypothesis concerning the structure and operation of ATP synthase and to Walker for his experimental confirmation of this model.

Observation of the Enzyme in Action
In 1997, final confirmation of the Boyer-Walker studies was accomplished by a research team at the Tokyo Institute of Technology and Keio University in Yokohama. The team was actually able to observe the rotation of the gamma subunit within the cylindrical a_3b_3 core of ATP synthase. They referred to the structure as "the smallest biological rotary motor known."

To study the behavior of the ATP synthase molecules, the Japanese researchers attached the a_3b_3 cylindrical subunit to a glass coverslip coated with Ni^{2+}-nitrilotriacetic acid (Ni-NTA). The Ni-NTA bonded strongly with molecules of the amino acid histidine that researchers had attached to the ends of the b subunits of the F_1 segment.

They then attached filaments of actin, which had been labeled with a fluorescent material, to the end of the gamma subunit opposite the glass coverslip. This step allowed them to use an epifluorescence microscope to observe any possible movement by the gamma subunit.

In their experiment, the research team actually forced ATP synthase to run in reverse. That is, they forced the enzyme to cause the hydrolysis of ATP to ADP. Their results are equally valid for the reverse reaction (ADP + P_i →ATP), however, because ATP synthase is a naturally reversible enzyme.

In their experiment, researchers measured the behavior of the ATP synthase molecule in the absence and presence of ATP. In the former instance, researchers observed no motion other than Brownian movement by the gamma subunit. When ATP was introduced to the experimental design, however, movement of the gamma subunit was easily observed. Of the 90 rotating filaments observed by the researchers, all rotated in an counterclockwise direction, when viewed from the membrane side of the structure. The direction of rotation exposed the three beta subunits in the sequence expected for the synthesis (or hydrolysis) of ATP.

Researchers noted that 44 of the 90 filaments continued to rotate for more than 1 minute, and some continued rotating for more than 10 minutes. The other filaments observed either tore loose from the gamma subunit or became immobile after a period of time.

To confirm their observations, researchers conducted a "blind" test on 1,800 filaments, some in the presence of ATP (ATP+), some in the absence of ATP (ATP–), and some in the presence of ATP and sodium azide (NaN_3) (ATP+; NaN_3+). Sodium azide is a known inhibitor of ATP synthase activity. They found that 25 filaments made more than 5 turns during 25 seconds of observation in the ATP+ sample, but no filament made no more than 2 turns during those 25 seconds in the ATP– and ATP+; NaN_3+ samples.

References

Brennan, Mairin, "Rotary Catalysis of ATP Imaged," *Chemical & Engineering News*, 31 March 1997, 27.

"Nano-Motoring," *Chemistry & Industry*, 7 April 1997, 244.

Noji, Hiroyuki, et al., "Direct Observation of the Rotation of F_1ATPase," *Nature*, 20 March 1997, 299–301.

Protein Folding

One of the most active fields of chemical research today deals with the folding of proteins. Proteins are large polypeptides that have complex three-dimensional structures. The primary structure of a protein is the sequence of amino acids that make up the protein. Today, discovering the primary structure of proteins has become a routine, if not exactly simple, task. It can be determined, for example, that a given protein consists of an arrangement of amino acids something like: GLY-GLY-TRP-ILE-GLU-CYS-ARG-GLY-PHE-VAL-LYS-. . . and so on.

Learning how to sequence the amino acids in a protein was an enormously important breakthrough. From another standpoint, however, knowing the primary structure of a protein is a trivial accomplishment. The functions performed by proteins, such as enzymes, hormones, and other essential biochemical compounds, are largely due to their three-dimensional structures. That is, the construction of a protein in a living cell does not end with the joining together of hundreds or thousands of amino acids. Instead, as the long polypeptide chain of a new protein is being put together, it begins to turn and fold to take on some characteristic and unique shape that allows the protein to perform its specific biochemical function. For decades chemists have asked themselves, What is it that determines the precise shape a protein will take? That is, what forces pull a polypeptide chain into an alpha helix, a beta sheet, or some other characteristic protein configuration?

Some fundamental elements of this process are well understood. For example, chemists have long known that hydrogen bonding, the formation of disulfide bonds and salt linkages, and van der Waal's forces are all involved in attracting one portion of a polypeptide chain to another portion and, thereby, pulling the chain into some characteristic three-dimensional shape.

But how does a polypeptide chain "know" where to form hydrogen bonds, disulfide linkages, or other forces of attraction among the seemingly unlimited number of possibilities that exist in a molecule containing hundreds of thousands of atoms?

Solving the puzzle of protein folding holds the potential for opening a variety of new doors in both pure and applied research. Once chemists learn the rules by which proteins "decide" how to fold, they may be able to replicate the construction of naturally occurring proteins or to design entirely new proteins with properties different from those found in nature.

During the last half of the 1990s, some critical steps were taken in solving the puzzle of protein folding. Although a thorough review of all the important discoveries cannot be provided in a book of this size, three examples of the kind of progress being made in this area are discussed below.

The Folding of Simple Proteins

A subset of the general problem described above deals with the nature of primitive proteins. The assumption is that the earliest proteins produced on Earth must have consisted of a relatively small number of amino acids. Over millions of years, the far more complex organic compounds now present on Earth slowly evolved. What is the simplest possible protein that one can imagine, then, that has the ability to form a three-dimensional structure and, presumably, could carry out essential biochemical functions?

In the early 1990s, that question was answered for one of the basic structures found in proteins, the alpha helix. In 1997, researchers at the University of Washington under the leadership of David Baker answered the same question for a second structure commonly found in proteins, the beta sheet.

For their research, Baker's team used a polypeptide segment made up of 57-amino-acid residues. This segment is found in many different proteins and contains 18 different amino acids. The team's goal was to find the smallest number of amino acids that could be used to substitute the 18 amino acids present in the natural product.

Researchers began with some obvious constraints. At least two different kinds of amino acids would be needed as substitutes, polar and nonpolar. Nonpolar amino acids are needed for the interior of the protein structure, and polar amino acids are needed for the exterior of the structure. Thus, the team began its search with isoleucine, which is nonpolar, and lysine and glutamate, which are both polar. Earlier research on simplified alpha helixes had suggested that as few as three different amino acids might be all that were needed to construct a viable polypeptide segment. However, Baker's team soon found that they required at least two more amino acids for their model. They selected alanine and glycine for these additions.

Testing the simplified substituted polypeptide segments was a challenging task. At each position in the naturally occurring segment, researchers had to choose among all possible choices and then test to see whether the altered segment behaved like the original naturally occurring segment. The test was used to see whether the altered segment would bind to another peptide in the same way that the naturally occurring segment did. If it did, the altered segment was regarded as a functional substitute for the original segment.

An additional observation made by the Baker team concerned the rate at which altered segments folded into the proper conformation. Researchers noted no significant difference between the rate for altered polypeptides and the rate for the naturally occurring segment. This result is important because it confirms that the development of more complex protein structures over time has *not* occurred as a result of evolutionary pressures to increase the rate efficiency of folding.

Baker's team concluded that a large number of viable polypeptide segments could be produced using no more than the five amino acids selected for the experiment. This result suggests that theoretical speculations about the origin of primitive proteins from a relatively small amino acid "alphabet" is reasonable.

An Abiotic Self-Organizing Oligomer

The folding that takes place in naturally occurring proteins represents the sum total of many different kinds of forces at work on the molecule. For example, hydrogen bonds, van der Waal forces, ionic attractions, and covalent bonding may all contribute to the precise type of folding that occurs within any one molecule. The structure and polarity of amino acids within the protein molecule also influence the direction in which amino acid residues face in the molecule and the way they interact with the surrounding medium. For example, amino acid residues with hydrophobic elements will tend to be repelled by water molecules in such a way as to contribute to the folding behavior of the protein molecule.

One way to study the complex nature of protein folding is to design simpler molecules in which fewer forces contribute to the folding process. In 1997, a research team at the University of Illinois reported on their study of *oligomers* with simpler structures than proteins that, nonetheless, demonstrate similar folding behavior.

Oligomer Folding

An *oligomer* is similar to a polymer in that it consists of a single monomeric unit repeated a number of times within the molecule. Compared to the hundreds or thousands of monomers present in a polymer, oligomers contain relatively few repeated units (no more than a few dozen or so). Oligomers are named according to the number of monomeric units present. For example, a dimer has two monomeric units; a trimer has three; a hexamer has six; and so on.

The oligomer selected by the Illinois team for study consisted of phenylacetylene units. These units consist of benzene rings containing ester units tied to each other by acetylene groups. A phenylacetylene dimer consists of two such groupings, a phenylacetylene trimer of three such groupings, a phenylacetylene tetramer of four such groupings, and so on.

These phenylacetylene oligomers are abiotic, that is, they do *not* occur in living organisms. Further, they lack the arrangements of atoms that make hydrogen bonds and most other types of bonding possible. Thus, the primary forces operating on these oligomers were expected to be the forces between the oligomers and the solvent in which they were dissolved.

The Illinois researchers calculated the theoretical free energy of a series of phenylacetylene oligomers in their linear and coiled states. They found that the longer the chain, the more likely the molecule would assume a helical shape.

They then synthesized 9 oligomers, with an even number of carbon atoms (the dimer, tetramer, hexamer, etc.), and studied their behaviors in chloroform and acetonitrile solutions. (See figure 1.9.) They found that oligomers with more than 8 monomeric units spontaneously formed helical coils and that those with more than 12 monomeric units did so most efficiently. Overall, as predicted by theory, the most efficient folding occurred with the 18-unit oligomer, the octadecamer.

The mechanism that drove this folding quite obviously appeared to be solvophobic (repulsion by solvent molecules) forces. Interestingly enough, the final product formed in this reaction was a molecule with a large, central cavity that would presumably be able to hold small molecules,

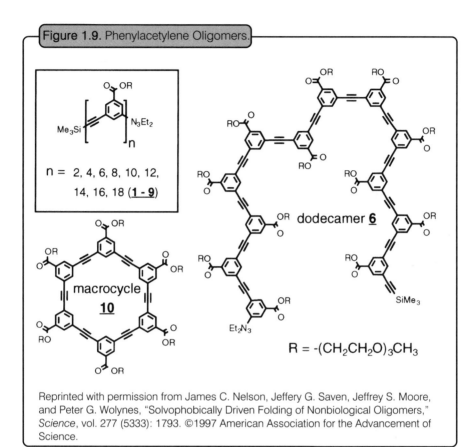

Figure 1.9. Phenylacetylene Oligomers.

n = 2, 4, 6, 8, 10, 12, 14, 16, 18 (1 - 9)

macrocycle 10

dodecamer 6

$R = -(CH_2CH_2O)_3CH_3$

Reprinted with permission from James C. Nelson, Jeffery G. Saven, Jeffrey S. Moore, and Peter G. Wolynes, "Solvophobically Driven Folding of Nonbiological Oligomers," *Science*, vol. 277 (5333): 1793. ©1997 American Association for the Advancement of Science.

ions, or other particles. Because of this property, the Illinois researchers envision possible future applications of such molecules, such as the catalysis of other reactions. These practical applications are probably farther into the future, however, than the improved understanding of the process of protein folding that their results already offer.

Designing Proteins

One of the ultimate goals of research on protein folding is to find ways to construct entirely new proteins that do not exist in nature. The demands and uses of such proteins are almost endless. Given the enormous variety of ways in which natural proteins are used both by organisms and by industry, the opportunity to design and build proteins with specific functions is likely to open a whole new expanse of research in chemistry and chemical engineering.

The task—to find a set of rules that determine how proteins fold—once seemed beyond comprehension. Research has moved forward so rapidly, however, that that objective has actually been attained. In 1997, B. I. Dahiyat, a graduate student in the Division of Chemistry and Chemical Engineering at the California Institute of Technology (Caltech), and S. L. Mayo, Professor of Biology at the Howard Hughes Medical Institute at Caltech, reported having developed a computer program that allows them to select a sequence of amino acids that can be used to manufacture an entirely new protein.

Redesign of the Zinc Finger

The goal of the Dahiyat-Mayo program was to find some set of amino acids that, when joined to each other, would form some preselected, three-dimensional conformation. The target structure they selected was that of a small polypeptide known as a *zinc finger,* which occurs naturally in DNA. The zinc finger was selected because it is a relatively small structure of only 28-amino-acid residues, yet it is capable of folding in such a way that the 3 basic structures of proteins are produced: an alpha helix, a beta sheet, and turning elements. In addition, the zinc finger is the smallest polypeptide structure known that is capable of folding without the involvement of disulfide bonds, metallic ions, or other subsidiary structures. Dahiyat and Mayo chose to aim for a zinc finger without the zinc atom that occurs in natural DNA.

The computational task of choosing among all possible combinations of amino acids to produce a 28-amino-acid residue is staggering. A total of 1.9×10^{27} such combinations exists. Obviously, nothing less than a high-speed computer could search through this number of options to find the correct possible combination of amino acid residues to be used in the primary structure of the polypeptide.

Dahiyat and Mayo's program was designed to analyze interactions between adjacent amino acid side chains and between amino acid side chains and the polypeptide backbone itself. The computer program was written to select among all possible combinations and interactions that would result in the lowest possible energy state. An additional factor included in the program was the fact that any one amino acid side chain can assume only a limited number of low-energy positions known as *rotamers*. Positions other than rotameric positions were discarded by the computer program.

When the presence of all possible rotamers was included with the number of all possible amino acid combinations, the total number of options the program had to consider was 1.1×10^{62}. The enormity of this task was simplified somewhat by including in the computer program a paradigm known as the dead-end elimination (DEE) theorem.

Results

Using the Dahiyat-Mayo theorem, a high-speed computer analyzed all possible rotameric configurations, for all possible amino acids, over a period of 90 hours. Ultimately, it produced a structure (called *full sequence design 1*, or *FSD-1*) that is remarkably similar to a naturally occurring zinc finger. Although the core of FSD-1 is somewhat larger than the core of the zinc finger, the overall configurations of the two structures are strikingly similar.

Somewhat surprisingly, the similarities in structure do not extend to the primary structures of the two molecules. Only 6 of the 28-amino-acid residues are identical, and only 11 are similar in the 2 structures. The authors point out that the relatively large difference in amino acid composition of the two structures "demonstrates the novelty of the FSD-1 sequence and underscores that no sequence information from any protein motif was used in [their] sequence scoring function" (Dahiyat 1997, p. 84).

The Dahiyat-Mayo algorithm represents an important step forward in understanding the mechanisms by which proteins fold. One reviewer stated that this work "will reveal much about how natural proteins adopt their folded conformations, while simultaneously allowing the design of entirely new polymers for applications ranging from catalysis to pharmaceuticals" (DeGrado 1997, p. 82).

References

Borman, Stu, "Proteins to Order," *Chemical & Engineering News*, 6 October 1997, 9–10.

Borman, Stu, "Synthetic Oligomer Mimics Protein Folding," *Chemical & Engineering News*, 22 September 1997, 8.

Dahiyat, Bassil I., and Stephen L. Mayo, "De Novo Protein Design: Fully Automated Sequence Selection," *Science*, 3 October 1997, 82–87.

DeGrado, William F., "Enhanced: Proteins from Scratch," *Science*, 3 October 1997, 82.

Nelson, James C., et al., "Solvophobially Driven Folding of Nonbiological Oligomers," *Science*, 19 September 1997, 1793–96.

Pennisi, Elizabeth, "Polymer Folds Just Like a Protein," *Science*, 19 September 1997, 1764.

"Proteins Encoded with a Smaller Amino Acid 'Alphabet'," *Chemical & Engineering News*, 6 October 1997, 25.

Riddle, David S., et al., "Functional Rapidly Folding Proteins from Simplified Amino Acid Sequences," *Nature Structural Biology*, October 1997, 805–09.

Tuchscherer, Gabriele, and Manfred Mutter, "Designing Novel Proteins," *Chemistry & Industry*, 4 August 1997, 597–600.

Wu, C., "An Alphabet for a Letter-Perfect Protein," *Science News*, 4 October 1997, 214.

Crystal Structure of a DNA Photolyase

Living organisms depend on the successful functioning of essential chemical compounds—especially nucleic acids and proteins—to remain healthy. It should not be surprising to discover, therefore, that organisms have evolved mechanisms by which these essential compounds can be repaired when they are damaged.

As an example of the damage that can occur to essential molecules, consider the effects of ionizing radiation. When ionizing radiation passes through an organism, it has the potential to strip electrons from molecules present in the organism. The loss of these electrons may result in the rupture of chemical bonds and the destruction of molecules. If these molecules are needed for essential biochemical reactions, the organism may become ill or even die.

Some concern is currently being expressed over the depletion of ozone in the Earth's atmosphere because of the health consequences ozone-depletion may have on organisms living on the Earth's surface. Ozone absorbs a significant fraction of the ultraviolet radiation reaching the Earth's outer atmosphere from the Sun. This ultraviolet light has a tendency to damage biochemical molecules by a process similar to the one described above. As levels of ozone in the stratosphere continue to decrease, we might expect to see an increased amount of damage to biochemical molecules and, as a result, an increasing level of damage to all forms of living organisms.

Photolyase as a Repair Agent

One of the most carefully studied repair systems in living organisms involves an enzyme known as photolyase. Various forms of photolyase occur in organisms as different as bacteria, archaebacteria, goldfish, rattlesnakes, and marsupials. The enzyme does not, however, occur in humans.

In 1995, scientists at the Howard Hughes Medical Institute of the University of Texas Southwestern Medical Center in Dallas announced that they had discovered the structure of the photolyase molecule found in *Escherichia coli (E. coli)*. (See figure 1.10.)

The photolyase molecule consists of 471 amino acids divided into 2 distinct sections, one consisting of both a-helixes and b-sheets, and the other consisting only of a-helixes. The two sections are joined by a loop containing 72 amino acids. The light-harvesting compound 5,10-methenyltetrahydrofolypolyglutamate (MTHF) is embedded in the crevice located between the 2 main portions of the molecule. The function of MTHF is explained below. A catalytic cofactor, flavin adenine dinucleotide (FAD), is located in the center of the helical section of the molecule. The function of FAD is also explained below.

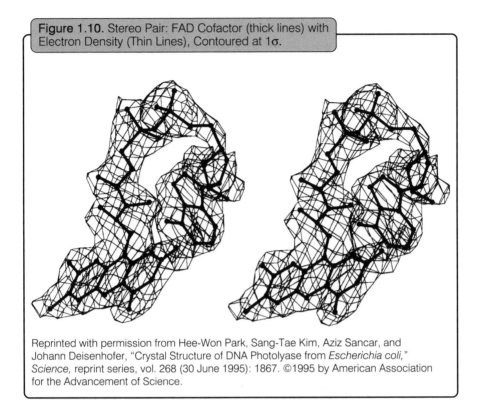

Figure 1.10. Stereo Pair: FAD Cofactor (thick lines) with Electron Density (Thin Lines), Contoured at 1σ.

Reprinted with permission from Hee-Won Park, Sang-Tae Kim, Aziz Sancar, and Johann Deisenhofer, "Crystal Structure of DNA Photolyase from *Escherichia coli,*" *Science,* reprint series, vol. 268 (30 June 1995): 1867. ©1995 by American Association for the Advancement of Science.

Function of Photolyase

The structure elucidated by the Howard Hughes group explains the mechanism by which photolyase repairs a DNA molecule. When such a molecule is damaged by ultraviolet radiation, chemical bonds in a DNA molecule break and then re-form to produce a mutant product known as cyclobutadipyrimidine (Pyr< >Pyr). Pyrimidine is a six-member heterocyclic ring found in the cytosine, uracil, and thymine components of nucleic acids. The formation of this product prevents a DNA molecule from either replicating or transcribing, essentially bringing about the death of the cell in which it is found.

Photolyase reverses the damage caused by ultraviolet radiation in a five-step sequence:

1. A photon of light energy is captured by the light-harvesting compound MTHF.
2. The additional energy provided by that photon is then transmitted to a FAD molecule, which releases an electron.
3. The electron, in turn, is transferred to the Pyr< >Pyr structure.

4. This causes a cleavage of two bonds in the Pyr<>Pyr ring. When those bonds break, the Pyr<>Pyr structure is broken, and normal pyrimidine rings are restored. The original structure of DNA is repaired and the molecule is able to function normally again.

5. The electron that initiated the rupture of the Pyr<>Pyr structure is returned to the FAD enzyme.

References

"DNA Repair Enzyme: A Structure Revealed," *Science News*, 8 July 1995, 20.
Park, Hee-Won, et al., "Crystal Structure of DNA Photolyase from *Escherichia coli*," *Science*, 30 June 1995, 1866–72.

BIOCHEMICAL REACTIONS

What is "life"? Over the centuries, humans have made many attempts to answer that question. Increasingly, a number of chemists are suggesting that "life" can be described to a large extent as a series of chemical reactions that take place within organisms. It is hardly surprising, then, that many chemists are now trying to better understand those reactions. The following selections represent this field of chemical research.

Vitamin K Function

In 1929, the Danish biochemist Carl Peter Nerik Dam discovered that a compound called vitamin K is essential for the clotting of blood in animals. In the absence of vitamin K, blood does not clot properly, and an animal can bleed to death. By contrast, an excess of vitamin K in the bloodstream can cause excessive clotting, leading to thrombosis (a blood clot) and the formation of embolisms (obstructions of blood vessels). The role of vitamin K in blood clotting is dramatized in the action of two drugs known as warfarin (once a common component of rat poisons) and dicumarol. Both compounds inhibit the action of vitamin K, thus preventing clotting and causing fatal hemorrhaging in animals.

Scientists have long known that the blood-clotting process involves a complex series of reactions sometimes referred to as the blood-clotting cascade (BCC). Essential to the BCC are a series of blood-clotting "factors," a variety of enzymes and proenzymes, vitamin K, and the Ca^{2+} ion. Exactly how all these species work in combination to produce clotting is not known. However, an important step in the BCC was elucidated in a report published in 1995 by Paul Dowd and his colleagues at the University of Pittsburgh.

Carboxylation in the Blood-Clotting Cascade

Dowd points out the role of carboxylation in the BCC. Carboxylation is the process by which a carboxylate ion (COO^-) is added to a molecule or ion. Carboxylation is repeatedly necessary at many stages of the BCC. For example, glutamate in Factors II, VII, IX, and X, and proteins C, S, Z, and other blood-clotting proteins must all be carboxylated for clotting to occur. In prothrombin (Factor II), 10 glutamates in a single region of the molecule must all be carboxylated before the molecule can be converted to thrombin.

Why is carboxylation so important in the BCC? Dowd suggests two major reasons, both related to the role of Ca^{2+} in clotting. First, the addition of a new carboxyl group in a molecule generally produces a second site of negative charge, which promotes binding with a Ca^{2+} ion:

Figure 1.11. Carboxylation in the Blood-Clotting Cascade, Part A.

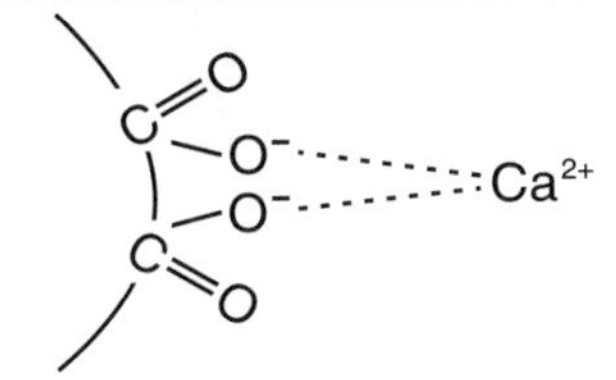

This arrangement allows calcium to act as a bridge between a blood-clotting factor and the surface of a cell membrane, a step that must occur before clotting can begin:

Figure 1.12. Carboxylation in the Blood-Clotting Cascade, Part B.

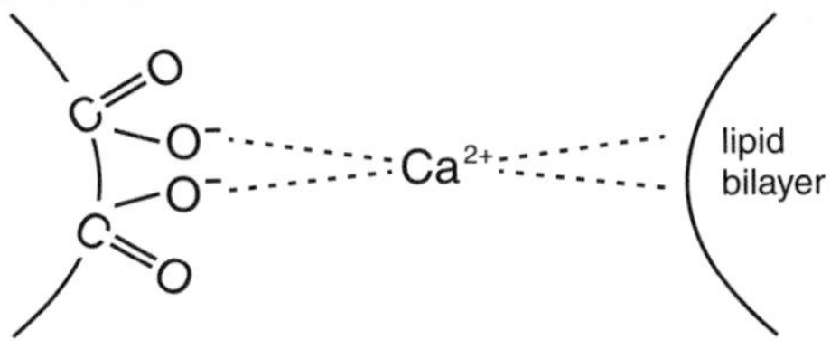

Second, the calcium ion can act as a bridge between two adjacent carboxylated protein molecules:

Figure 1.13. Carboxylation in the Blood-Clotting Cascade, Part C.

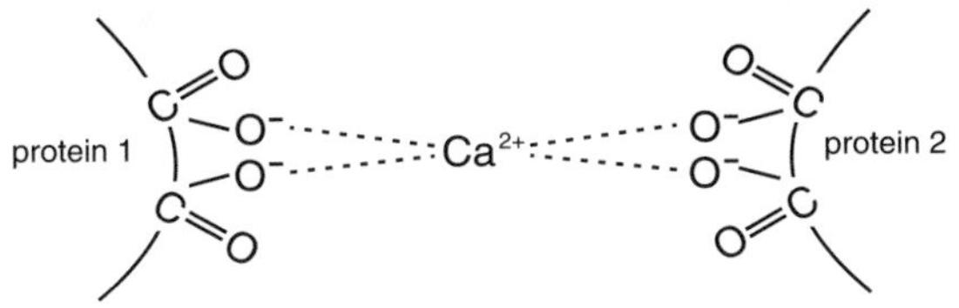

Dowd summarizes this effect as follows: "Binding to phospholipid membrane surfaces organizes the zymogens in the clotting sequence to promote cleavage of the proenzymes to their enzyme active forms" (p. 1684).

Role of Vitamin K

The above analysis clarifies the importance of carboxylation in the BCC. But what is the role of vitamin K in carboxylation? Dowd answers this question by reviewing steps involved in the so-called vitamin K cycle. (See figure 1.14.) In this cycle, vitamin K is first reduced to vitamin KH_2 by the action of the enzyme *reductase*, located on the reduced form of the energy-carrying molecule NAD(P)H. Vitamin KH_2 differs from vitamin K in that two double-bonded oxygens in the latter are converted to hydroxyl groups in the former.

In the next step of the vitamin K cycle, the enzyme carboxylase catalyzes the reaction between oxygen and vitamin KH_2 to form vitamin K oxide. Carbon dioxide formed in this reaction is then used to convert some other species—such as a protein—to its carboxylated form.

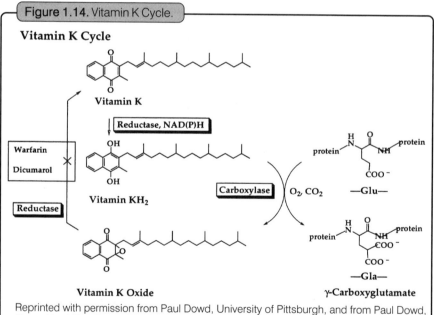

Figure 1.14. Vitamin K Cycle.

Vitamin K Cycle

Vitamin K

Reductase, NAD(P)H

Warfarin
Dicumarol

Reductase

Vitamin KH_2

Carboxylase O_2, CO_2

protein—Glu—

Vitamin K Oxide

protein—Gla—

γ-Carboxyglutamate

Finally, a reductase enzyme converts vitamin K oxide back to vitamin K, after which the cycle is ready to be repeated. It is at this point that warfarin and dicumarol exert their action on the clotting process. Either compound will combine with and inactivate a reductase enzyme or will otherwise interfere with the regeneration of vitamin K.

A significant portion of the Dowd paper on vitamin K action is devoted to a description of the team's analysis of the acid-base and energy aspects of the intermediary steps involved in vitamin K-related carboxylation. The paper outlines an "enzyme-free" experiment designed to test the mechanism by which the necessary pH and energy changes occur to make vitamin K-mediation carboxylation possible. That experiment elucidates in more detail the process by which changes in the vitamin K molecule make possible the carboxylation changes essential to the BCC process.

References
Dowd, Paul, et al., "Vitamin K and Energy Transduction: A Base Strength Amplification Mechanism," *Science*, 22 September 1995, 1684–90.
Lipkin, R., "How Vitamin K Helps Prevent Hemorrhaging," *Science News*, 23 September 1995, 199.

How Aspirin Works

A folk remedy for the relief of pain for more than 2,000 years has been to chew on the bark of a willow tree. Ancients knew that some component in willow bark had the ability to reduce pain and inflammation associated with headaches, arthritis, sore joints, and other ailments. It was not until the late nineteenth century, however, that the active agent in willow bark was identified. The German chemist Felix Hoffmann found that the compound known as *acetylsalicylic acid* was actually responsible for the pain- and inflammation-reduction qualities of willow bark. His employer, the Bayer division of I. G. Farber, soon marketed the sodium salt of this compound under the name *aspirin*.

The Role of Prostaglandins in Pain and Inflammation
Still, it was more than seven decades before scientists were able to find a chemical mechanism by which aspirin works. In 1971, an English pharmacologist, John Vane, was able to show that aspirin inhibits the production of prostaglandins. *Prostaglandins* are a group of biochemicals synthesized in the body from fatty acids, particularly from the fatty acid known as arachidonic acid. Prostaglandins have a number of functions in the body, including the regulation of blood pressure and coagulation, the rate of metabolism, glandular secretions, and contractions in the uterus.

The synthesis of prostaglandins in the body involves a complex series of chemical reactions. An important step in that series requires the participation of an enzyme known as *prostaglandin H_2 synthase (PGHS)*. PGHS is responsible for one of the first steps in which arachidonic acid is converted to a prostaglandin. In 1991, however, researchers at the University of California at Los Angeles (UCLA) and the University of Rochester found that PGHS exists in two forms, which they called *PGHS-1* and *PGHS-2*.

PGHS-1 is apparently present in low levels in a large variety of cell types, and it seems to have a variety of "housekeeping" responsibilities in cells. By contrast, PGHS-2 is produced only in cells that have been stimulated in some way or another, such as by injury. When PGHS-2 is produced, it causes pain, inflammation, and swelling. The function of aspirin, Vane's discovery seemed to suggest, was to inhibit the synthesis of PGHS-2. In the absence of PGHS-2, no prostaglandins are produced, and no pain or inflammation develops.

The Molecular Biology of Pain Suppression

In 1995, a research team under the direction of R. Michael Garavito, then at the University of Chicago, discovered the molecular mechanism by which aspirin inhibits PGHS and, thus, reduces pain and inflammation. The Garavito research represents an unusually fine example of the direction in which much biochemical research is directed today. Biochemists are no longer satisfied to say that aspirin (or sodium acetylsalicylate) inhibits the action of an enzyme known as PGHS. Instead, what they want to find out is how individual molecules of aspirin and PGHS combine with each other, fit together, or otherwise interact to produce this result. To answer such questions, the precise molecular structure of both substances must be determined.

What Garavito's team discovered was that PGHS consists of three independent units: the structure of one unit is similar to that of the epidermal growth factor (EGF); the second unit appears to provide a mechanism by which the molecule is attached to the surface of a cell membrane, and the third unit is the region in which catalysis actually takes place.

The third unit has a globular structure that consists of two adjacent but distinct sites. A different step in the conversion of arachidonic acid takes place at each of the sites. The active site of the enzyme consists of a long channel running through the middle of the molecule. Normally, a molecule of arachidonic acid enters the channel, attaches itself to binding sites within the channel, and undergoes a chemical change. This change is the first step in the conversion of arachidonic acid to prostaglandin.

Garavito and his colleagues studied the interaction between aspirin and the enzyme by using a modified form of the drug. They replaced one hydrogen atom in acetylsalicylic acid with a bromine atom to make the molecule easier to observe. They found that the modified aspirin reacted with an amino acid near the opening of the channel within the PGHS-1 molecule. This reaction resulted in the formation of a kind of gate within the channel. The gate completely closed the channel within molecules of PGHS-1, but only partially closed the channel within molecules of PGHS-2. That is, the gate prevented arachidonic acid from entering the enzyme active site on PGHS-1, inhibiting the formation of prostaglandins and the production of pain and inflammation.

The critical feature of this discovery, Garavito points out, is that we now have a structural explanation as to why aspirin is effective in inhibiting PGHS-1, but not PGHS-2. That information, in turn, can be used to develop other kinds of drugs that *are* more effective in inhibiting PGHS-2, a goal in which drug companies are, of course, very much interested. At the time of this writing, Garavito had begun research on the structure of PGHS-2 and the mechanism by which it might or might not be inactivated.

References

Burton, William, "Building a Better Aspirin," *The University of Chicago Magazine*, October 1995; <http://www2.uchicago.edu/alumni/alumni.mag/9510/October95Aspirin.html>.

Garavito, R. Michael, "The Cyclooxygenase-2 Structure: New Drugs for an Old Target?" *Nature Structural Biology*, November 1996, 897–901.

Lipkin, R., "Aspirin: How It Lessens Pain and Swelling," *Science News*, 12 August 1995, 102.

Loll, Patrick J., Daniel Picot, and R. Michael Garavito, "The Structural Basis of Aspirin Activity Inferred from the Crystal Structure of Inactivated Prostaglandin H_2 Synthase," *Nature Structural Biology*, August 1995, 637–43.

Picot, Daniel, Patrick J. Loll, and R. Michael Garavito, "The X-ray Crystal Structure of the Membrane Protein Prostaglandin H_2 Synthase-1," *Nature*, 20 January 1994, 243–49.

Transporting Metal Ions in Cells

Metal ions can pose difficult and serious problems for living cells. Such ions may be essential in small concentrations for the successful functioning of a cell. But in larger concentrations, those same ions may be toxic to the cell. Biochemists have long puzzled over how cells are able to handle potentially toxic ions. The cell must ensure that the ions are delivered to precisely the correct location within the cell in such a way that they do not cause damage to the cell during transit. One might draw a rough analogy by asking how a cancer specialist is able to obtain highly radioactive isotopes from a nuclear reactor and have them delivered safely to a hospital treatment room, without exposing large numbers of people to the dangerous isotopes.

Copper Chaperones in Yeast Cells

In 1997, an important breakthrough on the mechanism by which yeast cells transport copper ions was reported by a team of researchers at Northwestern University, the University of Michigan, and Johns Hopkins University. These researchers showed how copper ions are transported across the plasma membrane of a cell, through the cytoplasm, to their ultimate target organelle, a post-Golgi vesicle, where iron metabolism occurs.

The first step in this process involves the reduction of copper to copper(I) at the plasma membrane. The reduced copper is then transported across the membrane by means of the protein *Ctr1*, which has a special attraction for copper.

Scientists have found that Ctr1 passes copper(I) ions to one of three *chaperone protein molecules* once it has reached the inside of the cell membrane. These molecules are designated as *chaperones* because they have the responsibility of making sure that copper is delivered to exactly the right part of the cell. So far, three chaperone proteins have been identified: *Atx1*, which delivers copper to the post-Golgi vesicle; a *Lys7*, which delivers copper to an antioxidant enzyme in the cytosol, known as *Sod1*; and *Cox17*, which delivers copper to a cytochrome c oxidase within the mitochondria.

The Northwestern-Michigan-Johns Hopkins report has now clarified the details of the first of these transport processes. The main features of the Atx1 process involve the delivery of a copper(I) ion from the plasma membrane to a protein within the post-Golgi vesicle, known as *Ccc2*. Ccc2 then carries the copper(I) ion to an enzyme called Fet3, which is responsible for the uptake of iron in the vesicle.

Step-by-Step Transfer of Copper

The researchers identified two critical features in the movement of copper(I) from plasma membrane to post-Golgi vesicle. First, the transport of copper(I) can occur only when the two protein molecules, Atx1 and Ccc2, come into physical contact with each other. Once that happens, copper(I) can be physically transferred from Atx1 to Ccc2.

This transfer presents something of an energy problem for the cell, however. Atx1 must bind tightly to copper(I) at the plasma membrane, but then release the same ion relatively easily when it reaches Ccc2 in the target organelle. The mechanism by which this process seems to occur is as follows:

First, copper(I) bonds to two sulfur atoms in thiol groups within the Atx1 protein. Next, copper(I) forms a three-coordinate system in which it also bonds to a third sulfur from a thiol group in Ccc2. At this point, the copper(I) ion is bonded to two sulfurs in the delivering protein (Axt1) and to one sulfur in the receiving protein (Ccc2).

In the next step, one of the copper-sulfur bonds in Atx1 is lost. The copper(I) ion remains in a two-coordinate system, bonded to one sulfur in Atx1 and one sulfur in Ccc2. At this point, another change occurs, and a three-coordinate system is restored. In this system, however, two of those sulfurs are in the receiving protein (Ccc2) and only one is in the delivering protein (Atx1). Finally, the remaining bond between the copper(I) ion and an Atx1 sulfur is lost, and the copper(I) ion is held tightly within the receiving protein, Ccc2.

Significance

This research represents an important step forward, of course, in determining how cells handle potentially dangerous metallic ions. Additional studies with other chaperone proteins are now underway. One important application, however, is already obvious from these studies. The chaperone proteins Atx1, Cox17, and Lys7 found in yeast cells also have their counterparts in human cells. The human counterparts have either been shown to have, or are believed to have, a role in the development of disorders such as amyotrophic lateral sclerosis (ALS, or Lou Gehrig's disease), Menke's disease, and Wilson's disease.

References

Pufahl, R. A., et al., "Metal Ion Chaperone Function of the Soluble Cu(I) Receptor Atx1," *Science*, 31 October 1997, 853–56.

Rouhi, A. Maureen, "Escorting Metal Ions," *Chemical & Engineering News*, 10 November 1997, 34–35.

Valentine, Joan Selverstone, and Edith Butler Gralla, "Delivering Copper Inside Yeast and Human Cells," *Science*, 31 October 1997, 817–18.

A New Look at DNA Synthesis

One of the most dependable general rules in chemistry is not to depend completely on chemical laws. Over and over again in the history of chemistry, researchers have discovered that "firm rules" on which they thought they could depend are imprecise or inaccurate. In 1997, researchers at the University of Rochester, under the direction of Eric Kool, announced yet another example of the tendency of chemical "laws" to evolve and change.

Kool's research involved the process by which a DNA molecule makes copies of itself. For nearly four decades, chemists had thought that they understood that process rather well. Replication of a DNA molecule occurs when the double-stranded molecule comes apart, exposing two complementary chains. The enzyme DNA polymerase then begins to construct an exact copy of the original molecule using one of the two strands as a template for the procedure. That is, the enzyme selects nitrogen bases that "match up" with bases on the template. The rule it follows is that adenine and thymine bond to each other and cytosine and guanine bond to each other. Thus, when the enzyme encounters a thymine in the template, it searches for an adenine in the surrounding medium to bond with the thymine.

Chemists have long believed that the driving force between this matching process has been the way in which hydrogen bonds form between base pairs. An adenine molecule comfortably forms two hydrogen bonds with a thymine molecule, and a cytosine molecule comfortably forms three hydrogen bonds with a guanine molecule. The search for suitable hydrogen bonding, chemists have believed, drives the process by which DNA polymerase selects the correct nitrogen base to position in the new DNA strand.

Evidence for Steric Factors in Replication

Kool's research team tested the widely accepted hypothesis described above by altering the structure of a single strand of DNA and then using that strand as the template for the replication of the molecule. The change they made was to substitute difluorotoluene for thymine at the exact places where thymine would normally occur in the template strand. Difluorotoluene is similar in size and shape to thymine, but it lacks regions of partial positive or negative charge that would allow it to form hydrogen bonds with an adenine. The hypothesis behind Kool's work was that, lacking the ability to form hydrogen bonds, difluorotoluene would *not* be able to direct a correct synthesis of the corresponding strand of DNA.

In fact, the team obtained quite the opposite results. As DNA polymerase put together a new strand of DNA to correspond to the altered template strand, it "read" a difluorotoluene unit as being identical to a thymine unit. That is, it correctly selected an adenine to attach in position opposite each difluorotoluene unit, as it would have done had the position been occupied by a thymine. In fact, the rate at which replication occurred with an altered template containing difluorotoluene was the same as it was with the normal, thymine-containing template.

The conclusion to be drawn from this work is rather startling. It appears that it may not be the possibility of hydrogen bonding that DNA polymerase uses as a criterion for selecting nitrogen bases, but the suitability of shape and size of candidate molecules.

CHAPTER TWO
Chemical Technology Today

A dvances in basic and applied research and in technology are sometimes difficult to distinguish from each other. Chemists may make breakthrough discoveries while working in the field of basic research that have apparent applications in the real world. Should such discoveries be classified as advances in chemistry or in chemical technology? Answering this question can, at times, be difficult.

The studies reported in this chapter either arose out of efforts to apply chemical knowledge to the solutions of actual problems in everyday life or came from basic research but seem to have obvious applications in the real world. These studies can be subdivided into three general categories: health-related issues, materials science, and combating pollution.

HEALTH-RELATED ISSUES

Problems of disease and injury can often be regarded as chemical problems. For example, an organism may lack a chemical needed for normal biological function, or a chemical may not carry out its normal function in cells. Often the solution to a particular disease or disorder may involve supplying an organism with a chemical (a drug) that will provide a cure for or, at least, alleviate the symptoms of the ailment. The selections that follow illustrate research in which chemical principles have been used to resolve problems of injury, disease, or other health disorders.

Forty years of theory and research have not, of course, been thrown out the window. Indeed, other explanations of the Rochester team's research have been offered. It may be, for example, that difluorotoluene forms a new type of bond with adenine of which we are not yet aware. Additional research will be necessary before a more definitive answer can be provided about the newly reopened puzzle of DNA replication.

References

"Molecular Imposter Rebuts Long-Held Tenet of DNA Copying," <http://www.rochester.edu/pr/releases/chem/kool4.htm> (18 December 1997).

Moran, S., et al., "Difluorotoluene, a Nonpolar Iostere for Thymine, Codes Specifically and Efficiently for Adenine in DNA Replication," *Journal of the American Chemical Society*, 26 February 1997, 2056–60.

Wu, C., "Shape, Not Bonds, May Drive DNA Synthesis," *Science News*, 15 March 1997, 157; <http://38.214.184.12/sn_arc97/3_15_97/fob2.htm> (18 December 1997).

Enzyme Deficiency and Aging

Molecules in a living organism commonly undergo spontaneous changes. As an example, a proton from a hydroxyl group may shift to an adjacent or nearby carboxyl oxygen, forming an isomer of the original compound. Or a double bond may shift to a new position within a molecule, also producing an isomer. Or a characteristic functional group may move from one location in a geometric isomer to a second position, producing a racemic form of the first compound. When such changes occur in important biological molecules, profound, observable effects may occur within an organism. For example, if such changes occur in an enzyme, the altered form of the enzyme may no longer be able to carry out its original function, and an important biochemical reaction in the organism may no longer be able to occur. In addition, some scientists believe that one practical consequence of the accumulation of damaged proteins in an organism may be the onset of aging. Fortunately, organisms have ways to protect themselves against potentially disastrous changes such as these. Their bodies contain enzymes with the ability to repair damaged proteins. These enzymes can convert isomeric and racemic forms of a molecule, for example, back to their original functional forms.

Research

In 1997, scientists at the Universities of California at San Francisco and at Los Angeles, working under the direction of Edward Kim, performed a study on the ability of enzymes to repair damaged proteins. In this study, they used mice genetically engineered to lack a crucial repair enzyme. They investigated how the lack of this enzyme would affect development in the mice.

The study focused on various forms of a particular amino acid, aspartic acid, also known as asparagic or asparaginic acid. This acid occurs as the L-enantiomer in proteins in the form of L-asparagine and L-aspartate. These two forms of the acid are particularly unstable and are therefore subject to changes such as isomerization and racemization.

Earlier studies had showed that an enzyme known as L-isoaspartate (D-aspartate) O-methyltransferase is able to convert altered forms of asparaginyl and aspartyl residues to normal L-aspartyl residues in the test tube. The question that remained, however, was whether a similar reaction occurs in living organisms. That is, can the enzyme repair damaged proteins in living cells?

To answer this question, scientists first produced a genetically engineered mouse that lacked the *Pcmt-1* gene that codes for the production of L-isoaspartate(D-aspartate) O-methyltransferase. The researchers disabled the gene in the genetically engineered mouse, making it unable to manufacture the enzyme.

Results

Apart from being smaller and lighter in weight, the genetically engineered mice appeared to be similar to the control mice in nearly every respect during the first three weeks of their lives. Beginning on the 22nd day of the experiment, however, the engineered mice began to die off, and by the 60th day, all of the engineered mice had died. The mean and median day of death for these mice was 42 days.

Autopsies of the engineered mice found no obvious abnormalities that might have led to their early deaths. To better understand the cause of this early death syndrome, researchers videotaped the mice on a 24-hour schedule. What they observed was quite remarkable and unexpected. The videotapes showed that the mice suddenly went into seizures and died almost immediately from respiratory arrest.

Explanation

The violent nature of the mouse deaths suggested a problem with brain activity. As a result, the researchers examined brains of the dead mice for evidence of any abnormalities involving the suspected damaged proteins or the missing enzyme L-isoaspartate(D-aspartate) O-methyltransferase. They discovered that about 6% of the proteins in these brains were damaged, some of which might normally have been repaired by the missing enzyme. By comparison, less than 1% of the brain proteins in control mice of the same age were damaged. It was apparent that a connection between the missing enzyme L-isoaspartate(D-aspartate) O-methyltransferase and the ability to repair damaged proteins might exist.

Exactly how the damaged proteins were involved in the violent deaths of the engineered mice was not clear. The researchers proposed a number of possible connections, but they were unable to choose among them. An appealing hypothesis, however, appears to be that the damaged proteins might interfere with the transport of the neurotransmitter glutamate in the brain. The disruption of glutamate transport had previously been found to produce seizures similar to those observed in the engineered mice.

References

Brennan, T. V., et al., "Repair of Spontaneously Deamidated HPr Phosphocarrier Protein Catalyzed by the L-isoaspartate (D-aspartate) O-methyltransferase," *Journal of Biological Chemistry*, 7 October 1994, 24586–95.

Kim, Edward, et al., "Deficiency of a Protein-Repair Enzyme Results in the Accumulation of Altered Proteins, Retardation of Growth, and Fatal Seizures in Mice," *Proceedings of the National Academy of Science*, June 1997, 6132–37.

Travis, J., "Faulty Protein Repair Spurs Mouse Seizures," *Science News*, 14 June 1997, 365.

Stealth Erythrocytes

The function of an immune system in an animal is to protect it against invasion by some foreign substances and, thereby, to prevent disease and death. Although this system generally works to the advantage of the animal, it can also create problems for the animal. For example, when a human receives a transplanted heart, lung, kidney, or other organ from a donor, the recipient's body tends to react against and destroy the transplanted organ. Organ transplantation is always accompanied, therefore, by the problem of overcoming this immune reaction.

One instance in which the immune response can be especially troublesome is in the area of blood transfusions. Human red blood cells (RBCs or erythrocytes) can contain different antigens. The presence or absence of these antigens, which is determined genetically, produces different possible blood types, or blood groups. The presence or absence of an A or B antigen produces four possible blood types: A, B, AB, or O (which means that neither antigen is present). However, the presence or absence of another antigen, the Rh antigen, doubles the number of possible blood types, because each of the four blood groups can be either positive or negative for the Rh antigen. For example, the blood of a person with type A+ blood contains both the A antigen and the Rh antigen. This person, thus, can receive blood from a donor with type A+, A-, O+, or O- blood, without serious threat of an immune response. If the donor had type B blood in the previous example, however, the recipient would experience an immune response to the B antigen.

Fortunately, serious transfusion reactions are no longer a major problem because blood banks and transfusion centers routinely type the blood of donors and recipients to prevent an incorrect match between two individuals. However, transfusion reactions are more common in other situations. Individuals who receive frequent blood transfusions, such as those with thalassemia and sickle cell anemia, have an increasing likelihood of experiencing an immune reaction to the presence of several other minor blood group antigens. Eventually, it can become very difficult to find donors whose blood can be used for such frequent transfusions, without the danger of serious immune responses.

Disguising Red Blood Cells

Scientists have long searched for a solution to the RBC immune-response problem. One approach has been to develop synthetic blood, an artificial material that has the oxygen-carrying properties of erythrocytes, but fewer (or no) antigens associated with them. While some progress has been made in this direction, no product has yet been developed that meets the needs of a broad blood transfusion program.

Another approach has been to modify the surfaces of erythrocytes to remove the antigens that produce immune responses. Some success has been achieved with this technique for type B erythrocytes, but not for type A erythrocytes.

In 1997, a research team at the Albany Medical College, in Albany, New York, under the direction of Mark D. Scott, reported yet another approach to dealing with the problem of immune responses during blood transfusions. They bonded a plastic-like material, methoxypolyethylene glycol (mPEG) to the surfaces of mammalian erythrocytes. This material was used in an attempt to block RBC antigens that produce immune responses. The researchers then studied the properties of these erythrocytes and measured the extent to which they were rejected by other blood cells.

Methodology

Erythrocytes taken from humans, mice, and sheep were all treated with mPEG. The modified human cells thus produced were then tested with standard blood type antisera testing agents. The modified mouse cells were transfused into mice, where the modified cells' functions and survival times were measured. The modified sheep cells were similarly transfused into mice and studied. In addition, the modified sheep cells were exposed to human monocytes, (a type of white blood cell), which normally attack and destroy erythrocytes from foreign organisms.

Results

Modified human erythrocytes were studied under both light microscopes and scanning electron microscopes (SEMs) and were found to be morphologically similar to nonmodified erythrocytes. Modified erythrocytes were also found to have essentially the same ability to carry out osmosis and to transport oxygen as did nonmodified erythrocytes. That is, there appeared to be no functional or structural differences of biological significance between modified and nonmodified human erythrocytes.

The only difference observed was the one intended by the erythrocyte modifications: the cells did *not* clump when treated with antisera as they normally would in a blood type test. This result indicates that cells of the human immune system were not able to recognize the modified cells as foreign cells to be attacked and destroyed.

Interestingly enough, human monocytes also failed to recognize modified sheep erythrocytes. Under normal circumstances, such cells would have been attacked and destroyed by monocytes. Once "masked" by the presence of mPEG, however, the human monocytes did not recognize the sheep erythrocytes as foreign bodies.

In contrast to the *in vitro* experiments with human erythrocytes, the studies with modified mouse erythrocytes were conducted *in vivo*. In the first stage of these studies, modified mouse erythrocytes were transfused into mice, and the erythrocytes' survival times were measured. These erythrocytes had essentially the same survival times as nonmodified cells, which indicated that the mouse immune system also did not recognize the modified cells as foreign bodies. Moreover, this result was observed even after mice had been transfused many times.

A rather remarkable result was obtained when modified sheep erythrocytes were transfused into mice. Previous studies had shown that normal (nonmodified) sheep erythrocytes had a survival time of less than five minutes, as they were attacked and destroyed by mouse monocytes. When modified sheep erythrocytes were introduced, however, their survival times increased dramatically. The mouse immune system was apparently unable to distinguish between the modified sheep erythrocytes and the normal mouse erythrocytes.

Practical Applications

The researchers suggest that erythrocytes that have been modified with mPEG may solve the immune response problems that occur as the result of multiple transfusions and shortages of matching blood types. This experiment demonstrated that erythrocytes are not damaged in any way from mPEG bonding to their surfaces, and that the modified cells also function normally in the body. In addition, their manufacture is a relatively easy process, especially in comparison to some other techniques suggested for dealing with rejection responses during transfusions.

References

"Cloaked Blood Hides from Immune System," *Science News*, 9 August 1997, 92.

Scott, Mark D., et al., "Chemical Camouflage of Antigenic Determinants: Stealth Erythrocytes," *Proceedings of the National Academy of Science*, July 1997, 7566–71.

Bioengineered Analogs of Erythromycin

One of the great ongoing battles of modern science is the one fought between scientific researchers and pathogenic (disease causing) microorganisms. Scientists have been extraordinarily successful in finding compounds that will inactivate or kill such microorganisms, compounds such as penicillins, streptomycins, cephalosporins, tetracyclines, and sulfonamides. But the battle against disease is never over. Microorganisms are constantly mutating and evolving, developing new strains that are resistant to once potent medicines. Therefore, the discovery of new weapons against microorganisms is always an important breakthrough in chemical research.

Such a breakthrough was announced in 1997 by a research team headed by Chaitan Khosla at Stanford University. Khosla's research team found a method to commandeer the natural process by which an important antibiotic, erythromycin, is manufactured to produce a broad range of new and effective antibiotics.

Erythromycin Synthesis
The manufacture of erythromycin in cells is a complex and lengthy process involving more than two dozen distinct steps. The process can be compared to an automobile assembly line in which new parts are added to a car, one step at a time. In the case of erythromycin, individual units are added to the growing molecule, one step at a time, with each step requiring the participation of a specific enzyme. In cells, the individual units needed to sustain the process are obtained from the metabolic pool contained within the cell. Slight modifications in the units added to the growing erythromycin molecule account for modest differences in the final products. These differences in these products are indicated by their designated names *erythromycin A, erythromycin B, erythromycin C,* and so on.

The process of erythromycin synthesis can be generally represented as follows, where M_n is any one of the precursor units in the growing erythromycin chain, and E_n is the enzyme needed to bring about the next step in the chain.

$$M_1 + M_2\text{-}E_1 \to M_1M_2 + M_3\text{-}E_2 \to M_1M_2M_3 + M_4\text{-}E_3 \to \ldots + M_nE_{n-1} \to M_{erythromycin}$$

In some instances, the steps illustrated by this equation may actually occur at various sites within a given enzyme. In that case, the growing molecule is passed along from one active site to another, with an additional unit in the growing product being added at each site.

For some time, chemists have considered the possibility of making analogs available for natural substrates at any step in this process. The intention would be to present a cell with alternative paths to the production of erythromycin. These alternative paths would result in the formation of new forms of the antibiotic, different from any of the naturally occurring forms.

One could conceive, for example, that a cell might select a synthetic substrate, M, in step two of the above series, which would result in the formation of a product, M_1M_2M. This product would be similar to, but structurally somewhat different from, the compound $M_1M_2M_3$.

The problem is that, given the option, a cell prefers to use natural metabolites, rather than synthetic units, in the synthesis of erythromycin and other antibiotics. Thus, even if M is available in the cell, the natural metabolite M_3 is preferred in the synthetic process.

The Experiment

Khosla's team explored a new approach to the design suggested above. They altered a section of the gene that is responsible for the production of one of the enzymes (such as E_3) which, in turn, catalyzes the production of one of the intermediary products (such as $M_1M_2M_3M_4$). The gene alteration prevented the formation of the enzyme deoxyerythronolide B synthase (DEBS). DEBS is the enzyme responsible for the production of 6-deoxyerythronolide B (6-dEB), an early precursor of erythromycin.

Deactivation of DEBS, therefore, caused a total shutdown in the sequence of reactions leading to the production of erythromycin. As long as no 6-dEB was formed, the next step in the reaction could not occur, and the series of reactions came to a halt.

To ensure that the rest of the chain was operating properly, the researchers artificially added the missing 6-dEB to cells. After the addition, large amounts of erythromycin were found in these cells. This finding indicated that the enzymes were able to pick up from where the reaction had been cut off, and that the reactions leading to the final product still functioned normally.

In the final stage of their experiment, researchers added other compounds to cells containing altered genes. These compounds were structurally similar, but not identical, to 6-dEB. These analogs contained extra double bonds, hydroxyl groups, oxygen atoms, or other modifications that differed slightly from the natural component of the production chain.

What they found was that cells treated the analogs of 6-dEB just as they would have treated the natural precursor, 6-dEB, itself. The cells produced molecules that were structurally similar to erythromycin, but differed slightly from the natural product in some way. In fact, the cells seemed quite flexible in picking up and using substitutes for the missing natural precursor, 6-dEB, and going on to produce analogs of erythromycin.

The final test was to see whether these erythromycin analogs had similar biological properties to those of erythromycin itself. That is, did they function as antibiotics? The answer to that question was "yes." Researchers had, in fact, produced a host of new, synthetic products that could be used as antibiotics in situations essentially identical to those in which the natural product had been used.

Significance

The important point of this experiment is that scientists now have a new avenue for the manufacture of antibiotics. Other techniques for modifying the chemical structures of existing antibiotics, such as erythromycin, are available, but they tend to be difficult and clumsy. Indeed, the technique of "hijacking" a cell's normal biochemical mechanism for producing antibiotics to manufacture effective analogs is much simpler.

References

"Hijacking a Cell's Chemical Pathways," Academic Press Daily inScight, <http://www.europe.apnet.com/inscight/07181997/grapha.htm> (10 October 1997).

Jacobsen, John R., et al., "Precursor-Directed Biosynthesis of Erythromycin Analogs by an Engineered Polyketide Synthase," Science, 18 July 1997, 367–69; <http://www.sciencemag.org/cgi/content/fulltitleabstract; erythromycin!author1:jacobsen>.

Jensen, Mari N., "Science: New Tactics for Germ Warfare," The Seattle Times, 9 September 1997; <http://www.seattletimes.com/extra/browse/html97/altbact_090997.html>.

Salisbury, David F., "Discovery Offers Hope in Fight against Drug-Resistant Bacteria," Stanford Online Report, <http://www-leland.stanford.edu/dept/news/report/news/july30/antibiotics.html> (10 October 1997).

Nerve-Cell Growth

An important field of research in neurochemistry focuses on efforts to treat nerve cells that have been damaged by injury or disease. Such cells generally tend not to regenerate easily or well and, instead, tend to degenerate and die. When that happens, of course, important bodily functions may be lost. In 1997, two papers published in the Proceedings of the National Academy of Sciences reported on promising approaches to treating damaged nerve cells. One paper focused on regulating the growth of nerve cells, whereas the other focused on protecting nerve cells from destruction by free radicals.

Modified Fullerenes as Neuroprotective Agents

A study conducted at the Washington University School of Medicine in St. Louis examined the use of modified fullerenes as a mechanism for protecting nerve cells against attack by free radicals. Fullerenes are an allotropic form of carbon whose molecules consist of 60-carbon, soccer-ball-shaped spheres, or more complex modifications of such structures. They were discovered in 1985 by Harold Kroto, Richard E. Smalley, and Robert F. Curl, Jr., an accomplishment for which the three chemists were awarded the 1995 Nobel Prize for Chemistry.

The Washington University study was suggested because of two previously known facts about fullerenes and nerve cell damage. First, fullerenes had been found to be very reactive with free radicals. Second, free radicals had been implicated in nerve cell damage following a variety of central nervous system disorders, such as trauma, hypoxia-ischemia, and epilepsy. For example, earlier studies had shown that oxygen and nitric oxide free radicals are formed when certain neurotransmitter sites are overstimulated during disease or damage to the nervous system. Those free radicals are then known to attack and destroy otherwise healthy nerve cells.

Studies had also shown that the severity of free-radical attacks on nerve cells could be reduced by the use of certain "scavenger" molecules, although most such scavengers were not very effective in protecting nerve cells. The Washington team thus asked themselves whether the fullerenes would be more effective in accomplishing this task.

Methodology and Results
The first step in carrying out this experiment required a modification in fullerene molecules. Fullerenes tend to be soluble in organic liquids, such as toluene and benzene, and insoluble in water. To use fullerenes in a biological system, therefore, they had to be converted to a water soluble form. The Washington University researchers accomplished this objective by attaching three molecules of malonic acids to each fullerene molecule. They tried two forms of this modification: one in which all three addends were attached to the equator of the molecule, and one in which all three addends were attached to one of the two hemispheres of the molecule.

In the next step of the experiment, researchers measured the effectiveness of modified fullerene to react with free radicals, such as the hydroxyl radical (OH^-) and the superoxide anion (O_2^-). They found that fullerenes were anywhere from 10 to 100 times as effective at free radical scavenging as were other compounds used for that purpose. The researchers concluded that the modified fullerenes are "the only class of antioxidant compounds that [they] have worked with to date that can fully block intense, rapidly triggered, NMDA receptor-mediated excitotoxicity in [their] cortical neuronal cultures" (Dugan 1997, p. 9437). (NMDA is the abbreviation for N-methyl-D-aspartate, a compound that causes nerve cell death.)

In the final step of this experiment, researchers implanted mini-pumps containing modified fullerenes into mice genetically altered to carry the gene for familial amyotrophic lateral sclerosis (FALS). Their goal was to discover whether the modified fullerenes were as effective *in vivo* as they had appeared to be *in vitro*.

Researchers found that both forms of the modified fullerene reduced the symptoms of FALS in mice and extended their lives. In control mice, the first symptoms of FALS began to show up at about 8 weeks of age. Mice treated with a modified fullerene, however, did not show such symptoms until about 10 days later, on average. In addition, control mice died at an age of about 2 months as a result of the disease, whereas treated mice survived an average of 6 to 12 days longer.

A significant difference in the effectiveness of the two modifications of fullerene was observed. The form in which all three malonic acid groups were attached to one hemisphere of the molecule was found to be significantly more effective in delaying symptoms of FALS and extending a mouse's life than was the form in which all three groups were attached to the equator of the molecule. Later studies suggested to the researchers that this difference may be due to the greater efficiency of the first form in penetrating the lipid bilayer contained in cell walls.

The researchers concluded from this study that modified fullerenes may have potential benefit in the treatment of nerve damage resulting from disease and injury and that a modification that is soluble in both water and organic liquids (such as the "all-in-one-hemisphere" form) may be the most effective form for use.

Treating Nerve Damage

Surgeons have developed sophisticated techniques for the treatment of nerve cells damaged by disease or injury. The most common approach is to make nerve grafts on a damaged nerve cell using neuronal material taken from elsewhere in the patient's body. This procedure, while useful, has some drawbacks. One problem is that the process may require a series of surgical procedures. A second difficulty is that the region from which neuronal material is taken may then become numb.

In a paper published in the same issue of *Proceedings of the National Academy of Sciences,* scientists at the Massachusetts Institute of Technology (MIT) and at Children's Hospital of Harvard University in Boston described a new technique for encouraging the regeneration of damaged nerve cells.

The purpose of this research was to assess the usefulness of an electrically conductive polymer in promoting the growth of nerve cells that have been damaged by injury or disease. Similar studies have been conducted in the past using materials such as veins and muscles from the affected individual and prosthetic tubes made from collagen, laminin, fibronectin, silicone, and a variety of other polymers.

One disadvantage of most of these materials is that they are electrical insulators. Yet, scientists have long known that electrical impulses tend to promote the growth and regeneration of nerve cells, although the reasons for that phenomenon are still not understood. The MIT and Children's

Hospital researchers decided to study the use of a conductive polymer in the construction of a growth-generating material.

Two decades ago, the term *conductive polymer* would usually have been considered a contradiction in terms. The vast majority of polymers known then were electrical insulators, incapable of carrying an electric current. In the early 1970s, however, chemists found that certain polymers can be made to be conductive and that the conductivity of those polymers is very much a function of *dopants* (impurities intentionally added to the polymer).

Methodology and Results. Researchers chose to use an oxidized form of the polymer polypyrrole (PP) because of its abilities to conduct an electric current and to maintain a positive charge spread equally across its outer surface. The material had other assets as well, namely, that it could be wetted easily and that it was biologically compatible with mammalian cells.

Special rat cells (PC-12 cells) were laid out on a surface made of a thick PP film that was laminated onto a supporting structure made of a second polymer, poly(lactic acid-co-glycolic acid). The surface was then treated with nerve growth factor (NGF) and subjected to an electric potential of 100 mV for 2 hours. A similar control plate of PC-12 cells on PP film were similarly treated with NGF, but were not subjected to an electric potential. Changes in cells for both the experimental and the control plates were examined by means of phase-contrast optical microscopes.

Cells exposed to an electric potential were found to show more pronounced growth than control cells not exposed to electric potential. Mean neurite length for experimental cells was 18.14 μm compared to 9.5 μm for control cells. Thus, the use of a conductive polymer with an applied electric potential improved nerve cell regeneration by roughly 100%.

Researchers suggest two advantages in using conductive polymers for nerve cell regeneration. First, the electrical flow is restricted to the surface on which nerve cells rest rather than being spread out generally by an external electrical field outside of the plastic. Second, the conductive plastic can be fabricated to have almost any electrical properties needed for a particular application.

References

Dugan, Laura L., "Carboxyfullerenes as Neuroprotective Agents," *Proceedings of the National Academy of Sciences*, August 1997, 9434–39.

Schmidt, Christine E., et al., "Stimulation of Neurite Outgrowth Using an Electrically Conducting Polymer," *Proceedings of the National Academy of Sciences*, August 1997, 8948–53.

Watanabe, Dave, "Buckyballs Fight Nerve Damage," *Daily Science News* (Washington University School of Medicine), 19 August 1997; <http://www.helios.org/news/37.html>.

Wu, C. "Polymer, Buckyballs Combat Nerve Damage," *Science News*, 23 August 1997, 119.

A Possible Iron-Transporter Protein

As is the case with most minerals, an organism's body must maintain a concentration of iron within a relatively narrow range. A reduced level of iron tends to lead to anemia, a condition characterized by loss of energy and weakness that can result in death. According to some estimates, between 500 million and a billion people worldwide suffer from some form of anemia.

On the other hand, an excess of iron can also be a problem. The condition known as *hemochromatosis,* for example, can lead to inflammation of the liver, diabetes mellitus, and heart failure.

Scientists have been searching for over three decades for the protein responsible for transporting iron throughout the body. The discovery of this protein could promise relief to the millions of people who suffer from an excess or deficiency of iron in their bodies.

A possible candidate for the iron-transporter protein was identified in a pair of unrelated studies in 1997. One study, led by Nancy C. Andrews, at the Howard Hughes Medical Institute at Children's Hospital and Harvard Medical School in Boston, examined a strain of mice with a hereditary form of anemia known as *microcytic (mk) anemia.* Microcytic anemia is a condition in which red blood cells are of unusually small size. These mice tend to be smaller in size and weight than normal mice. They also tend to develop skin lesions that appear shortly after they are born. Mk mice lack the ability to absorb iron through the intestinal wall and to utilize the iron they do take in.

Gene Mutation

Andrews's research team found the probable cause of anemia in the mice they studied to be caused by a mutation in a gene known as *Nramp2* located on mouse chromosome 15. This mutation makes it impossible for the mouse to produce a protein that appears to be responsible for the uptake and transport of iron in the body.

Confirmation of the Andrews study was reported simultaneously by a second research group, under the direction of Matthias A. Hediger at Brigham and Women's Hospital in Boston. In the Hediger study, rats were fed a diet low in iron. The assumption was that, in such a circumstance, a rat's body would begin to increase the rate of iron-transporter protein production. This increase would provide the rats a way of avoiding iron-deficiency disorders that would otherwise develop because of the diet low in iron.

At the conclusion of the first stage of this experiment, Hediger's group made copies of RNA taken from the iron-starved rats and injected those copies into frog eggs. Finally, they immersed those eggs in a solution rich in iron. They looked for the egg cells that absorbed iron from the solution most efficiently. It would be the injected RNA in those eggs, they reasoned, that would be responsible for the production of iron-transport protein. The gene identified in the Hediger experiment was, in fact, the same one discovered by Andrews's group, *Nramp2*.

Both research teams noted and commented on the relationship between the suspect gene, *Nramp2*, and a homolog, *Nramp1*. *Nramp1* codes for the formation of a protein involved in an animal's resistance to bacterial attack. It may be, Andrews has suggested, that the *Nramp1* protein functions to withhold iron from bacteria, thereby disrupting the bacteria's life functions and providing the rats the resistance associated with the protein.

Significance

Identification of the iron transport protein could have enormous practical benefits. On the one hand, people with hemochromatosis might be able to take a drug that would selectively inhibit the action of the iron transport protein. This therapy would be much more pleasant than current methods of treating the disorder, which involve removal of blood from the patient. It might also be possible to find ways to promote the action of the iron transport protein in people with various forms of anemia. That kind of therapy would have very widespread benefits for the millions of people whose bodies do not naturally absorb or process iron adequately.

References

Fleming, Mark D., et al., "Microcytic Anaemia Mice Have a Mutation in *Nramp2*, a Candidate Iron Transporter Gene," *Nature Genetics*, 16 August 1997, 383–86.

"Iron Transport Protein Found," <http://www.traders.co.uk/insulintrust/iron.htm> (26 September 1997).

Travis, J., "A Protein that Helps the Body Pump Iron," *Science News*, 2 August 1997, 68.

Rules for the Design of New Drugs

The discovery of antibiotics was one of the great breakthroughs in the history of medicine. Antibiotics are substances produced by living organisms that weaken or kill microorganisms. Many antibiotics belong to a family of organic compounds known as polyketides. Polyketides contain a backbone of at least a dozen carbon atoms to which are attached two or more carbonyl groups. The chain also includes at least one cyclic or aromatic group. The polyketides also include a number of important nonantibiotic drugs and medicines, such as FK506 (an immunosuppressant), doxorubicin (an anticancer agent), and monensin and avermectin (antiparasitics).

Traditionally, polyketides with therapeutic value have been discovered empirically, by trial and error. Scientists have long wished, however, that they could better understand the biochemical mechanisms by which microorganisms produce polyketides, so that such compounds could be generated by more rational means. The development of genetic engineering techniques has made possible the realization of that dream. In 1995, a team of researchers at Stanford University and the John Innes Centre in Norwich, U.K., used genetic engineering techniques to outline a scheme for the rational design of polyketides.

Natural Polyketide Synthesis

A long series of experiments has elucidated the general steps in the process by which an organism manufactures a polyketide. The key to this process is a group of enzymes known as *polyketide synthases (PKSs)*. PKSs catalyze a variety of reactions that result in the formation of a polyketide. One of these reactions is the condensation of two acyl thioesters. Acyl thioesters consist of carboxylic acid fragments (such as the acetyl group in acetic acid) joined to a sulfur atom. The condensation of two acetyl thioesters, for example, results in the formation of a four-carbon chain, while the condensation of two propionyl thioesters yields a six-carbon chain. Condensation reactions can be repeated a number of times until chains of 16, 18, 20, 24, or more carbon atoms are produced.

A second reaction catalyzed by PKSs is the reduction of a keto group. After condensation, the extended chain of carbon atoms contains $n \div 2$ carbonyl groups, where n is the number of carbons in the chain. Any one or more of these carbonyl groups can be reduced to a hydroxyl group by a PKS.

PKSs can also cause the cyclization and aromatization in the growing carbon chain. Cyclization simply involves the closing of one part of a chain to yield a ring within the chain, while aromatization involves the introduction of double bonds in the ring to produce a benzene-like structure.

The final products of polyketide synthesis differ from each other on the basis of chain length, degree of carbonyl group reduction, and number and location of ring and aromatic groups.

Natural and Synthetic Polyketides

As noted above, the polyketides available for therapeutic uses have traditionally come from trial-and-error experiments on natural products manufactured by microorganisms. Understanding the mechanisms by which such compounds are produced, however, makes possible a more rational approach to the synthesis of artificial polyketides.

This approach involves the identification of specific portions of PKSs that catalyze various chain length, reduction, cyclization, and aromatization reactions and then locating the genes responsible for the production of these active sites. Finally, bacteria can be engineered to contain genes that code for any given combination of properties desired for a polyketide. Such a scheme is outlined in figure 2.1.

To test the validity of this scheme, the Stanford–John Innes group designed two entirely new polyketides, which they called *SEK43* and *SEK26*. The two new compounds were produced by specifying genes that would produce compounds of a specific chain length, in which specific carbonyl groups had been reduced, and that contained ring and aromatic groups in specific locations.

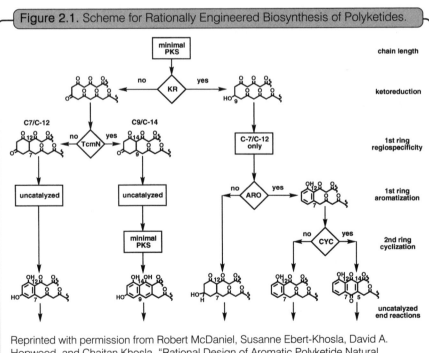

Figure 2.1. Scheme for Rationally Engineered Biosynthesis of Polyketides.

Reprinted with permission from Robert McDaniel, Susanne Ebert-Khosla, David A. Hopwood, and Chaitan Khosla, "Rational Design of Aromatic Polyketide Natural Products by Recombinant Assembly of Enzymatic Subunits." *Nature*, vol. 375 (15 June 1995): 553. ©1995 Macmillan Magazines Limited.

Researchers concluded their report by pointing out that other chain modifications should be possible. These modifications would include the transfer of groups from one part of a molecule to another and further oxidation or reduction reactions. With so many synthetic options available, they conclude that it should be possible "to generate libraries of 'unnatural' natural products with structural diversity that is comparable to that observed in nature" (McDaniel 1995, p. 553).

References

Lipkin, R., "New 'Design Rules' Yield Novel Drugs," *Science News*, 17 June 1995, 374.

McDaniel, Robert, et al., "Rational Design of Aromatic Polyketide Natural Products by Recombinant Assembly of Enzymatic Subunits," *Nature*, 15 June 1995, 549–54.

Genome of an Ulcer-Causing Bacterium

One of the great ongoing debates in the history of science has to do with the ultimate nature of the phenomenon we call "life." Are living organisms nothing other than a highly complex combination of chemical compounds that obey clearly defined chemical and physical laws? Or does "life" involve the intervention of forces beyond those that can be studied and understood by scientific methods?

A large and intriguing body of research bearing on this question focuses on the identification of the genomes of various organisms, ranging from those of the simplest bacteria to the human genome itself. A major breakthrough in this research was announced in 1997 when a team of 42 researchers at 6 institutions reported on the complete genome of the bacterium *Helicobacter pylori*. This bacterium was found in 1984 to be responsible for gastric inflammation and peptic ulcers, conditions once thought to be due to excess stomach acid caused by stress and other factors. Scientists estimate that almost half of the world's population harbors this bacterium, making it, according to some experts, "probably the most common bacterial infection of humans" (Tomb 1997, p. 539). Thus, a better understanding of the nature of *H. pylori* potentially has enormous practical significance, leading to the possible development of drugs and vaccines for the treatments of various stomach disorders.

Results of the Research

The *H. pylori* genome was found to consist of 1,667,867 base pairs arranged in a single, circular chromosome. The genome also contains 1,590 coding sequences, which make up 91% of the chromosome and are responsible for specific properties and functions of the bacterium. The structure of the genome is shown in figure 2.2.

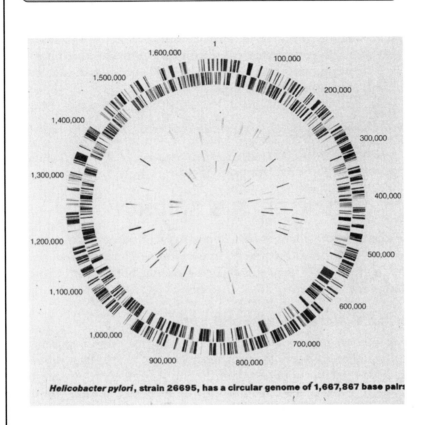

Figure 2.2. Circular Representation of *H. pylori* 26695 Chromosome.

Helicobacter pylori, strain 26695, has a circular genome of 1,667,867 base pairs

Reprinted with permission from Jean-F. Tomb et al. "The Complete Genome Sequence of the Gastric Pathogen *Helicobacter pylori*," *Nature,* vol. 388 (7 August 1997): 541. ©1997 Macmillan Magazines Limited.

Researchers were able to correlate specific regions of the bacterial genome with specific properties and functions of the organism, such as cell division; protein secretion; recombination, repair, and restriction systems; transcription and translation; adhesion and development of genetic variation; virulence; regulation of gene expression; and metabolism. For example, they were able to identify the genes responsible for the formation of two proteins, a cytotoxin-associated protein and an active vacuolating cytotoxin, responsible for the damage caused by the bacterium to host epithelial cells. Those proteins are responsible for the formation of acid vacuoles in cells, bringing about their destruction and the development of ulcerlike symptoms in the host.

The genome also reflects the presence of a large number of genes responsible for the production of proteins used in the manufacture of the bacterial membrane. Researchers suggest that this feature of the genome may explain adaptations made by the bacterium that allow it to survive in the otherwise harsh environment of the human stomach, which has a pH of about 2. The complete gene sequence can be viewed on the Internet at <http://www.tigr.org/tdb/mdb/hpdb/hdpd.html>.

References

"Complete Genome of Ulcer-Causing Bacterium Obtained," *Chemical & Engineering News*, 11 August 1997, 35.

Tomb, Jean-F., et al., "The Complete Genome Sequence of the Gastric Pathogen *Helicobacter pylori*," *Nature*, 7 August 1997, 539–47.

MATERIALS SCIENCE

One of the oldest challenges for chemists is to find new ways of using familiar materials or, alternatively, to develop new materials to be used for new applications. The selections below outline both kinds of research reported in recent years.

Spherical Fullerenes as Solid Lubricants

Most car owners understand the importance of proper vehicle mainte-nance. Parts that are lubricated regularly last far longer than those that are not. This principle applies more generally, of course, to any mechani-cal system in which two components rub or slide against each other. The development of efficient lubricants, therefore, has long been an impor-tant field of research in chemical engineering.

The vast majority of lubricants are liquids, similar to the motor oil used in passenger cars and trucks. But liquid lubricants are unsatisfactory for certain applications, such as in devices that operate in a vacuum or in space. For such applications, solid lubricants must be used. In 1997, scientists working at the Center for Technological Education–Holon and the Weizmann Institute, both in Israel, reported on the development of fullerene-like nanoparticles that perform far more efficiently than many commercially available solid lubricants.

Solid Lubricants in Bulk and Nanoparticle Form

The materials studied by the Israeli scientists were made of tungsten disulfide (WS_2). Powdered WS_2 is already used as a solid lubricant in a variety of industrial operations. The compound appears to function as a lubricant because it exists on the molecular scale in the form of thin sheets that slide back and forth over each other easily. The same property is found in graphite, one of the most commonly used of all solid lubricants.

Tungsten disulfide has its drawbacks as a lubricant, however. Over time, it becomes sticky and gummy, losing its efficiency as a lubricant. Analysis suggests that this change takes place when chemical changes occur along the edges of WS_2 sheets. "Dangling" chemical bonds at the ends of sheets appear to react with oxygen and water, forming tungsten trioxide (WO_3). As this product accumulates, WS_2 sheets become clogged up and lose their ability to slide over each other easily.

In the Israeli research, a new form of WS_2 was prepared: hollow, multi-layered nanospheres. Researchers found that when WS_2 sheets are heated to high temperatures, they tend to curl up and form spheres in much the same way that sheets of carbon form fullerene molecules. The major difference between these two reactions, however, is that the carbon change results in the formation of discrete fullerene molecules, while the WS_2 change results in the formation of larger, nano-sized particles up to 120 nm in diameter. These particles are hollow, like fullerene molecules, but they consist of a number of concentric layers.

Results

Mechanical tests of the hollow WS_2 nanoballs showed them to be significantly superior in both function and wear compared to WS_2 powder. Furthermore, the nanoballs are slightly compressible, so they can absorb an increased load placed on them more easily than can rigid WS_2 powders. Finally, WS_2 nanoballs can be expected to last much longer as lubricants than can WS_2 powder for two reasons. First, outer layers of a nanoball may wear away, but additional layers beneath them remain in place. Second, nanoballs have no ends, as do sheets of WS_2. Thus, nanoballs are unlikely to have dangling bonds that can react with oxygen, water, and other materials in the environment.

Practical Applications

In their report, the Israeli researchers outlined the possible practical value of their studies. The starting materials for the production of hollow WS_2 nanoballs are readily available, they point out, so that "the cost of the material is therefore not expected to prohibit large-scale applications in the automotive and aerospace industry, which currently use the apparently inferior dichalcogenides" (Rapoport 1997, p. 793).

References

Rapoport, L, et al., "Hollow Nanoparticles of WS_2 as Potential Solid-State Lubricants," *Nature*, 19 June 1997, 791–93.

"Tiny Balls are Tough but Slippery," *Chemistry & Industry*, 7 July 1997, 498.

Safer Superconducting Materials

The 1987 Nobel Prize for Physics was awarded to J. Georg Bednorz and K. Alex Müller for their discovery of a material that becomes superconductive at 35 K (–238°C). This award was significant because the Bednorz-Müller discovery had been announced only a year earlier. Traditionally, Nobel Prizes are given for discoveries made many years earlier, often as much as 20–40 years previously.

The Bednorz-Müller Nobel Prize recognized an amazing breakthrough in the development of superconducting materials. Over a period of more than seven decades, since the discovery of superconductivity by Heike Kamerlingh-Onnes in 1911, only very modest progress had been made in finding materials that become superconductive at temperatures much above absolute zero. Bednorz and Müller totally revolutionized research in this field with their discovery of a material with a dramatically higher transition temperature (the temperature at which a material becomes superconductive).

One ongoing problem with this research has remained, however. At least one of the components of all high-temperature superconducting materials—usually thallium or mercury—is toxic to humans. Therefore, such materials must be prepared, under carefully controlled conditions, and their applications are somewhat limited by their toxicities at high temperatures.

Structure of a Nontoxic High-Temperature Superconducting Material

In 1997, Ching-Wu (Paul) Chu and his colleagues at the Texas Center for Superconductivity at the University of Houston reported on the development of a new superconducting material containing no toxic elements. Chu and his colleagues have long been at the forefront in research on superconducting materials and hold most of the records for high-temperature superconducting materials. The nontoxic material they produced has the general formula $Ba_2Ca_{n-1}Cu_nO_x$ and has a transition temperature of 126 K. The compound itself is unlikely to find applications in superconducting appliances since it is unstable in moist air—adding carbonate and hydroxide ions, which alter the structure of the material. The modified form of the material, however, still has a transition temperature of 90 K.

The primary significance of the Chu team's latest discovery is probably not the specific compound discovered, but the structural features to which it may owe both its superconductivity and its lack of toxicity. Traditionally, high-temperature superconducting materials have complex crystal structures consisting of layers of discrete species. A compound with the general formula of $A_mE_2R_{n-1}Cu_nO_{2n+m+2}$, for example, has a unit cell containing m layers of AO inserted between two layers of EO on top of n layers of CuO_2 interleaved by n-1 layers of R, where A, E, and R are various cations. An example of such a compound has the formula $CuBa_2YCu_2O_7$ (or $YBa_2Cu_3O_7$).

These layers are arranged in two large domains, called the *charge-reserve block* and the *active block*. (See figure 2.3.) The charge-reserve block has the general structure (EO)(AO) $\cdots$ (AO)(EO), and the active block has the general structure $(CuO_2)(R)(CuO_2)$ $\cdots$ $(R)(CuO_2)$. Theory suggests that charges produced in the charge-reserve block are transferred to the active block, the region in which superconductivity actually originates.

The Chu team investigated the possibility of introducing calcium into a barium-copper precursor with the formula $Ba_2Cu_3O_{5+x}$, with $0 \leq x \leq 1$. They expected that calcium would be inserted between BaO layers in the charge-reserve block to form a compound with the putative structure of $Ba_2Ca_{n-1+x}Cu_{n+y}O_z$, where $0 \leq y \leq 1$. The product of this reaction was found to have a transition temperature of 126 K.

Figure 2.3. The Proposed Schematic Atomic Arrangement of Interstitially Doped 0223-Ba, Excluding O the Charge-Reservoir Block.

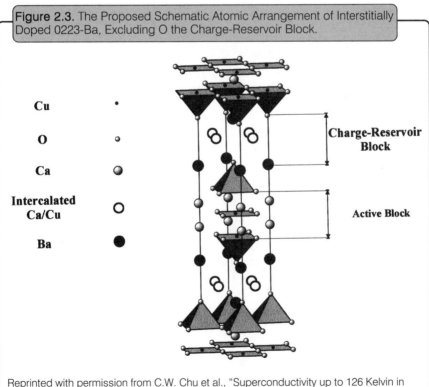

Cu   •

O   ○

Ca   ◔

Intercalated Ca/Cu   ○

Ba   ●

Charge-Reservoir Block

Active Block

Reprinted with permission from C.W. Chu et al., "Superconductivity up to 126 Kelvin in Interstitially Doped $Ba_2 Ca_{n-1} Cu_n O_x$ [02(n-1)n-Ba]," *Science*, vol. 277 (5329): 1081. ©1995 American Association for the Advancement of Science.

The exact structure of the new material has yet to be determined since it degrades when irradiated with the beam of a transmission electron microscope, the instrument usually used to determine such structures. Researchers feel confident enough to say, however, that the material they produced "represents a genuine new compound family" (Chu 1997, p. 1083).

References

Chu, C. W., et al., "Superconductivity up to 126 Kelvin in Interstitially Doped $Ba_2Ca_{n-1}Cu_nO_x$ [02$(n$-1$)n$-Ba]," *Science*, 22 August 1997, 1081–83.

"Superconductor Flies High without Volatile Toxic Ingredients," *Chemical & Engineering News*, 25 August 1997, 26.

Soluble Pigments

Color chemists distinguish between two terms, *dye* and *pigment*. Both terms refer to a compound used to add color to paints, plastics, inks, and other products. The difference is that dyes are soluble in the carrying medium, while pigments are not. The insolubility of pigments means that special efforts—often requiring considerable time, energy, and expense—must be made to prepare emulsions of pigments before they can be used.

In 1997, chemists at the Pigments Division of Ciba Specialty Chemicals in Basel, Switzerland, announced a new method for dealing with this problem. The method involves the synthesis of a pigment precursor, called a *latent pigment*, that is soluble in solvents in which the pigment itself is insoluble.

Theory and Application

A common reason for the insolubility of pigments is the existence of hydrogen bonds between adjacent molecules in the colorant. If these bonds could be broken, the Ciba researchers reasoned, the modified pigments might become more soluble. One way to break the hydrogen bonds, of course, would be to replace polar hydrogen atoms in molecules with other nonbonding groups.

The compound used to test this hypothesis was 3,6-diphenyl-1,4-diketopyrrolo[3,4-*c*]pyrrole, generally known as *DPP*. Pairs of hydrogen bonds between adjacent DPP molecules create a tightly bound network that accounts for the low solubility of this compound in solvents such as xylene (less than 10^{-3} g/L).

The Ciba team treated DPP with di-(*t*-butyl)-dicarbonate in tetrahydrofuran at room temperature. In the reaction that followed, amino hydrogens in the DPP molecule were replaced by *t*-butoxycarbonyl [$(CH_3)_3$–C–O–CH=O] groups incapable of forming hydrogen bonds. The modified DPP was much more soluble in xylene (36 g/L) than was the original form of the compound. Therefore, it could be used and applied very easily, in much the way that dyes are used and applied.

The modified DPP was then exposed to temperatures of about 180°C. At these temperatures, the modified DPP lost its added groups, regained hydrogen atoms, and was restored to its original pigment form. The byproducts of the thermal conversion were carbon dioxide and isobutane, both gases that can easily be removed from the pigmented material. Researchers noted that the pigment formed by thermal decomposition was "identical to . . . DPP prepared by standard synthetic procedures" (Zambounis 1997, p. 131).

This discovery has potentially profound implications for the pigment industry. If it is shown to be applicable to pigments other than DPP, very significant cost reductions may be achievable in the manufacture and use of organic pigments.

References
"Soluble Pigments? Here's How," *Chemical & Engineering News*, 14 July 1997, 36.
Zambounis, J. S., A. Hao, and A. Iqbal, "Latent Pigments Activated by Heat," *Nature*, 10 July 1997, 131.

Molecules That Trap the Energy of Light

Arguably the most important single set of chemical reactions on Earth are those that occur during the process of photosynthesis. In those reactions, the energy of sunlight is converted to chemical energy that makes possible the production of complex carbohydrates from carbon dioxide and water. Scientists still do not understand all of the details of photosynthesis, although important progress in unraveling that scheme has been made in recent decades.

Scientists have long been interested in developing synthetic systems that mimic the mechanism by which solar energy is converted to chemical energy naturally in green plants. Success in such efforts would make it possible to better utilize our single most important source of energy—the Sun—for everyday, commercial applications.

One of the problems encountered in the early stages of this research was the complexity of systems required for the conversion of solar to chemical energy. Photosynthesis itself requires more than a dozen reactions for the conversion of carbon dioxide and water to carbohydrates, a system that would be far too inefficient in human-made systems for commercial applications.

Supramolecular Trimetallic Complexes
In the mid-1990s, a number of studies reported success in the construction of individual molecules with the capability of converting the energy

of photons into the production of electron pairs. The assumption is that those electron pairs would then be available to carry out chemical changes. Some of the most interesting of this work has been done in the laboratory of Karen J. Brewer at Virginia Polytechnic Institute and the State University at Blacksburg.

Brewer's work has focused on the creation of very large molecules (supramolecules) consisting of three metals with a variety of ligands and other bridging groups. An essential component of these molecules has been the compound $Ru(bpy)_3$, where *bpy* is an abbreviation for the compound 2,2'-bipyridine. Scientists have long known that this compound has the ability to capture a photon and to use that energy to produce an excited electron, which can then be transferred to a substrate. Although studies have been made of this reaction as a possible mechanism for the production of solar-generated electrical energy, the reaction has a number of disadvantages and is generally too inefficient for commercial applications.

Thus, Brewer's team investigated the possibility of incorporating a light-absorbing unit and an electron-collecting unit within a single molecule. The goal was to increase the number of excited electrons produced per molecule and, hence, the efficiency of the overall reaction.

Methodology and Results
To achieve this goal, Brewer's team prepared a number of compounds containing a variety of noble metals and complex organic groups. The general formula for one group of compounds, for example, was $\{[(bpy)_2Ru(BL)]_2MCl_2\}^{n+}$, where *M* represents Ir^{3+}, Os^{2+}, and Rh^{3+}; *BL* represents 2,3-bis(2-pyridyl)pyrazine, 2,3-bis(2-pyridyl)quinoxaline, or 2,3-bis(2-pyridyl)benzoquinoxaline; and *bpy* represents 2,2'-bipyridine. In these large molecules, two ruthenium complexes are joined to a central, third metal complex, in which the third metal is iridium, osmium, rhodium, or ruthenium.

When exposed to light, one of these molecules undergoes a photochemical reaction in which the ruthenium complex first absorbs a photon of light. The energy of the photon is then used to raise an electron in the attached ligand to an excited state. Finally, the excited electron is transferred to the electron collector, the central core complex containing the third metal atom.

The intriguing feature of this process is that each trimetallic supramolecule contains two ruthenium complexes and can, therefore, generate two electrons, an efficiency twice that of earlier compounds studied for this purpose. Thus, the Brewer work suggests a mechanism, by which synthetic, inorganic molecules can be used to generate electrical charges, which is similar to mechanisms used by green plants to convert solar energy to chemical energy. Those electrical charges could conceivably be used directly in the generation of electrical power (as in solar cells) or to drive chemical reactions, such as the splitting of water molecules to produce hydrogen and oxygen.

References

Brauns, Eric, et al., "Electrochemical, Spectroscopic, and Spectroelectrochemical Properties of Synthetically Useful Supramolecular Light Absorbers Mixed with Polyazine Bridging Ligands," *Inorganic Chemistry*, 18 June 1997, 2861–67.

Lipkin, R., "New Molecules Harness the Energy of Light," *Science News*, 6 April 1996, 212.

Milkevitch, Matthew, et al., "A New Class of Supramolecular, Mixed-Metal DNA-Binding Agents: The Interaction of Ru^{II}, Pt^{II} and Os^{II}, Pt^{II} Bimetallic Complexes with DNA," *Inorganic Chemistry*, 24 September 1997, 4534–38.

Research on Flexible Ceramics

Ceramics are among the oldest structural materials used by humans. Early cultures learned how to use naturally occurring materials, such as clay and other earthy materials, to make a variety of products ranging from bricks, tile, terra cotta, and decorative jewelry to large buildings, brick walls, and aqueducts. Ceramics continue to be widely popular for industrial applications such as refractory ovens, spark plugs, Portland cement, and ceramic fibers. The popularity of ceramics is based largely on their hardness, stiffness, and resistance to high temperature and attack by chemical agents. Indeed, many ceramics are superior with regard to these properties than more popular construction materials such as metals and plastics. The one serious drawback of ceramics, however, is their brittleness. When exposed to stress, they tend to break easily, rather than simply bending, as do metals and plastics.

A New Flexible Ceramic

A flood of research is now underway to find a way of adding the one property that ceramics lack—flexibility—to those they already possess. An important step in that direction was announced in 1997 by a research team under the direction of Yoshiharu Waku at the Ube Research Laboratories in Ube City, Japan. The research team produced a new ceramic composite consisting of single crystals of alumina (Al_2O_3) and gadolinium aluminum oxide (gadolinium aluminum pervoskite, $GdAlO_3$, or GAP). This material can be bent, without fracture, at temperatures up to 1,873 K.

The new composite was formed by a process known as *unidirectional solidification,* in which a material is forced by the shape of its container to solidify very slowly, primarily in a single direction. The Waku team had earlier used this technique to make a composite ceramic consisting of alumina and yttrium aluminum oxide (yttrium aluminum garnet, $Y_3Al_5O_{12}$, or YAG), which retained its flexibility at 2,073 K, just below its melting point of about 2,100 K.

When the product of the fabrication of the latest ceramic was studied, it was found to consist of two discrete crystals, one of Al_2O_3 and one of $GdAlO_3$, intermixed with each other. Researchers described the composite as having "a microstructure consisting of three-dimensionally continuous and complexly entangled single-crystal Al_2O_3 and single-crystal GAP" (Waku 1997, p. 49). They compared this kind of structure to those found in liquid crystals, water-oil-soap mixtures, and copolymers.

Stress tests conducted at 1,873 K showed the new ceramic to be nearly twice as ductile as the Al_2O_3–YAG composite. The latter material experienced a dislocation of about 0.2 mm, under a stress of 400 MPa, before fracturing. By contrast, stresses of up to 700 MPa caused complete flexibility in the Al_2O_3-GAP composite, which never fractured under the conditions used in the experiment.

Practical Applications

The practical applications of this kind of ceramic appear to be endless. These ceramics appear to have all of the best properties of metals and plastics, in addition to having many more of their own. As examples of the possible applications of this kind of material, the Ube researchers mentioned the manufacture of gas-turbine and power-generation systems that have to operate at very high temperatures, with no means for cooling.

References

"Flexible Ceramic Takes the Heat," *Chemistry & Industry*, 15 September 1997, 711.
Wilson, Elizabeth, "New Ceramic Bends Instead of Breaking," *Chemical & Engineering News*, 8 September 1997, 12.
Waku, Y., et al., "A Ductile Ceramic Eutectic Composite with High Strength at 1,873 K," *Nature*, 4 September 1997, 49–52.

A Porous Silicon Biosensor

A number of circumstances exist in which scientists would like to be able to detect the presence of certain organic compounds in a sample. They might be searching for the presence of a drug, a particular type of DNA, or a natural biochemical, such as a vitamin, in a living system. A wide variety of devices have been developed for this purpose. These devices, generally known as *biosensors*, can detect and transmit information about any property of a living system. In its simplest form, a biosensor consists of a source of energy (such as a beam of light), an object to be examined, and a device for detecting and interpreting the energy reflected from the object.

In 1997, researchers at the Scripps Research Institute and the Skaggs Institute for Chemical Biology in La Jolla, California, announced the development of a biosensing device with significantly improved properties. The researchers worked under the direction of M. Reza Ghadiri, Associate Professor of Chemistry at Scripps. The Ghadiri biosensor uses ordinary white light as the source of energy, porous silicon as the platform on which objects are to be examined, and a charge-coupled device as a mechanism for analyzing light reflected off the porous silicon.

Porous Silicon
The new biosensor is made of a form of silicon that has only recently been developed and studied. It is called *porous silicon* because its surface is etched with many deep grooves. The grooves range in depth from 1 to 5 μm and have diameters up to 200 nm. Under an electron microscope, the surface of porous silicon looks like an eroded landscape with untold numbers of silicon pillars jutting up from the material's base.

When ordinary light is shined on the surface of porous silicon, an interference pattern is produced similar to that produced when light is shined on a diffraction grating. A diffraction grating is a piece of glass, plastic, metal, or other material in which hundreds or thousands of very thin lines have been etched in close proximity to each other. Light reflects off the surface of a diffraction grating (or porous silicon) in such a way that wave fronts combine with each other either constructively or destructively. The result is an interference pattern with dark and bright rings that represent destructive and constructive interference.

Porous Silicon as a Biosensor
The Ghadiri team found that it could attach a variety of molecules (substrates) to the surface of porous silicon. It appears that these molecules bond to the sides of the silicon pillars that make up the silicon surface. The common feature of the substrate molecules used in this experiment was that they were all able to "recognize" some other molecule. For example, in one experiment, the team used a piece of DNA, 16 nucleotides long, as a substrate. A DNA sequence has the ability to "recognize" and bond to a second segment of DNA whose nitrogenenous base sequence is exactly complementary to that of the original DNA.

When white light was shined on porous silicon containing recognition molecules, it produced different interference patterns depending on whether a second molecule had bonded to the substrate or not. For example, one might designate the interference pattern of white light shined on porous silicon to which a sample of DNA is bonded as "A." If a complementary strand of DNA is then introduced to the porous silicon, it will bond with the substrate DNA strand and form an interference pattern that might be called "B." When a noncomplementary strand of DNA is introduced to the porous silicon—one that will *not* bond to the substrate DNA—there is no change in the interference pattern produced.

Striking Properties of the Biosensor
A number of properties make this kind of biosensor of special interest. First, Ghadiri's team noticed a "cascade-sensing" effect. An example of this effect is as follows:

Substrate molecule A recognizes and binds to molecule B, and the biosensor reports on the presence of molecule B. Molecule B, however, is also a recognition molecule that can bind to another molecule, called molecule C. Thus, the system consists of molecule A bound to B bound to C. The cascade-sensing effect means that the biosensor recognizes the presence of C as well as the presence of B. This effect can be repeated a number of times, with the biosensor recognizing the presence of each new molecule that binds to the existing system.

A second advantage of the porous silicon biosensor is its sensitivity. Under the best conditions used in their experiments, the researchers were able to detect as little as a few femtograms (10^{-15} g) of material.

Finally, the biosensor is a relatively simple device that can be transported and used easily. It is also small enough to fit into the palm of one's hand. These properties make it promising for use as a biosensor in a wide variety of everyday applications.

References
Lin, Victor, S. Y., et al., "A Porous Silicon-Based Optical Interferometric Biosensor," *Science*, 31 October 1997, 840–43.
Wu, C., "Biosensors Respond with Colored Light," *Science News*, 15 November 1997, 317.
Zurer, Pamela, "Porous Silicon Biosensor," *Chemical & Engineering News*, 15 September 1997, 7.

Development of a "Copper Chip"

Ever since the development of transistors more than three decades ago, aluminum has been used for wiring connecting elements within a semiconductor chip. Aluminum is used in spite of the fact that it does *not* conduct electricity as well as other common metals, particularly silver and copper. Silver is, of course, too expensive to use in chips, but why not use copper? Most large-scale electrical wiring consists of copper because of its superior conductive properties.

A number of reasons explain the traditional use of aluminum rather than copper in chips. First, copper is much more difficult to work with than aluminum at the small dimensions needed in the manufacture of chips. Second, finding a dependable way to affix copper to the semiconductor substrate of silicon has been difficult. By contrast, aluminum can be attached relatively easily to a silicon base. Finally, copper has a somewhat greater tendency to react with the semiconductor itself than does aluminum.

The Need for Greater Speeds Produces Solutions

Chip technology, like all forms of technology, is driven by marketplace demands. Manufacturers continually require smaller, faster, and more powerful computing devices in their electronic instruments. A fundamental limit created by the use of aluminum in a chip is the frequency with which electrons can travel through the metal, a limit of about 400 MHz. This limit has forced electronic engineers to find ways of overcoming the inherent problems in using copper in the manufacture of chips.

In 1997, both IBM and Motorola announced that they had found ways to solve these problems and that they would soon be producing chips with copper wiring. A key to solving this problem was the development of a new metallurgical technique for laying copper down on a silicon wafer so that it would remain in place and not react chemically with the surrounding material. The use of copper rather than aluminum promises to raise the speed at which electrons can travel through a circuit to approximately 1,000 MHz.

The dimensions of the new copper circuits are of the order of 0.20 micron, smaller than anything so far produced in the industry. By contrast, a human hair has a diameter about 500 times that of a single copper wire on a circuit. The new copper circuits also have an effective length of no more than 0.12 micron, the shortest current path yet developed in a chip. The length of this path will make it possible to significantly increase the speed at which electrical messages can be sent through a chip and, hence, the speed at which electronic devices can operate.

By reducing the dimensions of wiring on the chip, manufacturers expect to be able to greatly increase the number of transistors installed on a single chip. At the present time, the chips in desktop processors hold between 5.5 and 7.5 million transistors. With the new technology, that number should increase to 50–100 million transistors per chip.

For consumers, the development of a copper chip should mean that everyday devices, such as cellular telephones and personal computers, should become smaller, more convenient, and faster to use.

References
Davis, Jim, "Motorola Has a Copper Chip Too," <http://www.news.com/News/Item/0,4,14777,00.html> (20 October 1997).

"IBM Discovers Way to Make Copper Chip," *San Francisco Chronicle*, 22 September 1997, A2.

"IBM Makes a Copper Chip," <http://www.the-view.com/news/0997/2503.html> (30 September 1997).

Wu, C., "Computer Chips Take a Leap Forward," *Science News*, 27 September 1997, 196.

Supercritical Carbon Dioxide

The simple process of dissolving substances is an enormously important step in most industrial operations. For example, decaffeinated coffee is produced by dissolving caffeine from coffee beans. The usual process of decaffeination makes use of water to dissolve caffeine. But this process is not very efficient because caffeine is not very soluble in water. Researchers have been looking for other solvents and methods to make the decaffeination of coffee less expensive.

Water is, of course, by far the most important single solvent. Its ability to dissolve a very wide range of substances has long given it the name of the "universal solvent." The problem is that most organic compounds are not soluble in water, and many important industrial operations require solutions of organic substances. As a result, chemists have developed, over time, a variety of solvents for organic substances. Among the most popular organic solvents have been benzene, chloroform, carbon disulfide, carbon tetrachloride, and organic families such as the ethers, alcohols, and halogenated hydrocarbons.

Although these solvents were once widely popular, many have come under question because of health and environmental issues. All of the individual compounds mentioned above have been shown to be carcinogenic, teratogenic, or to cause other potential health hazards. The halogenated hydrocarbons, for example, have been found to have deleterious effects on the Earth's atmosphere. Thus, many of the more popular commercial solvents for organic reactions are now either banned or in limited use.

In their search for substitutes for traditional industrial solvents, chemists have turned their attention to supercritical carbon dioxide ($scCO_2$). Supercritical carbon dioxide is formed when gaseous carbon dioxide is heated to temperatures in excess of $31°C$ at pressures greater than 73.8 atm. Under these conditions, the gas takes on the properties of both a gas and a liquid. It has the ability to dissolve substances, as do solvents like water and benzene, but it also has the ability to pass through porous surfaces, as it does in the gaseous phase.

Supercritical carbon dioxide has been used for nearly three decades in a wide variety of industrial operations. Applications have included the removal of caffeine from coffee and nicotine from tobacco, the extraction of essential oils from plants for use in perfumes and cosmetics, the cleaning of metals in electronics, and the production of polymers. In the latter case, methods have been developed for dissolving monomers and catalysts in $scCO_2$ as the first step in polymerization reactions.

The health and environmental advantages of using $scCO_2$ in industrial operations are obvious. Carbon dioxide is an important constituent of the atmosphere, and it is involved in a variety of biochemical processes. In the amounts that would be used for the purposes above, its intentional or accidental release to the atmosphere would be expected to have no deleterious effects on plants, animals, or the environment in general.

Supercritical Carbon Dioxide and Aqueous Reactions

One of the fundamental drawbacks in the use of $scCO_2$ in industrial operations is that polar and ionic substances are generally not soluble in the material. In 1996 and 1997, reports began to appear describing methods by which this drawback could be overcome. The 1996 report described work by a research team consisting of scientists at the University of Texas at Austin, the University of Nottingham in England, the State University of New York at Buffalo, and the University of Colorado at Boulder. The leader of the team was chemical engineer Keith P. Johnston at Texas.

The approach developed by Johnston's team was to form a microemulsion by adding water and an emulsifying agent to $scCO_2$. The goal was to produce an emulsion somewhat similar to that obtained when soap is added to a mixture of water and oil. The emulsifying agent used by the Johnston team was ammonium carboxylate perfluoropolyether (PFPE). The team conducted a series of tests to determine whether a microemulsion had, in fact, been formed. They were eventually convinced that such a microemulsion had been formed and that it contained "some type of aggregated water domain" (Johnston 1996, p. 624) that might serve as a solvent for polar and/or ionic substances.

To test this hypothesis, an effort was made to dissolve a simple protein, the bovine serum albumin protein. They selected this protein because of its moderate size and because it could be studied easily by fluorescence spectrometry. The team found that the protein did, indeed, dissolve in the water-$scCO_2$ emulsion.

The Nature of the Water-$scCO_2$ Emulsion

The 1996 report did not discuss in detail the nature of the water-$scCO_2$ emulsion. That question was the topic of a second report published a year later by members of the original team from Nottingham and Texas. This team demonstrated the existence of particles they called *reverse micelles* in which a typical micelle encompassed a sphere of water they referred to as *bulk water*. (See figure 2.4.) That is, the reverse micelle consisted of a tiny core of pure water surrounded by a film of "interfacial" water to which were bonded the hydrophilic ends of surfactant molecules of PFPE. Researchers were able to show that the bulk water in reverse micelles exhibited solution properties similar to those of pure water itself.

Figure 2.4. Schematic Representation of the Water Environments in Reverse Micelles or Microemulsions Showing the Possible Environments in which Water May Be Found.

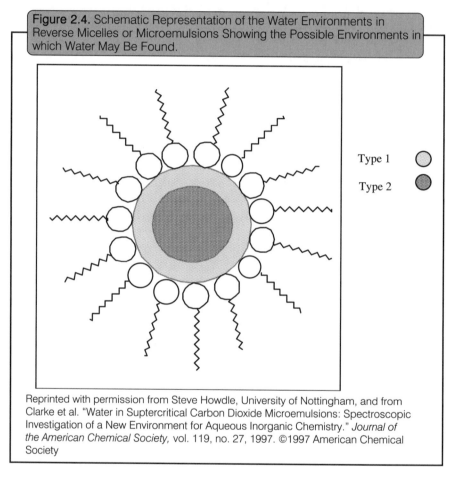

Type 1

Type 2

Reprinted with permission from Steve Howdle, University of Nottingham, and from Clarke et al. "Water in Suptercritical Carbon Dioxide Microemulsions: Spectroscopic Investigation of a New Environment for Aqueous Inorganic Chemistry." *Journal of the American Chemical Society,* vol. 119, no. 27, 1997. ©1997 American Chemical Society

For example, when potassium permanganate was added to a water-$scCO_2$ emulsion, the complete sample turned purple, a color that all chemistry students associate with the presence of permanganate ion. The appearance of the purple color demonstrated that the emulsion had dissolved $KMnO_4$ because $scCO_2$ by itself does *not* dissolve the salt.

The emulsion's solution properties were also demonstrated by the team's ability to carry out a standard chemical reaction within it. First, an acidic solution of potassium dichromate was added to the water-scCO$_2$ emulsion. The appearance of a characteristic orange color confirmed that the salt had been dissolved. Next, sulfur dioxide gas was passed through the emulsion and a color change from orange to bluish-green was observed. This familiar test for the presence of sulfur dioxide gas indicated that dichromate ion was converted to chromium(III) ion.

Johnston's team explained the properties of the water-scCO$_2$ emulsion as being the result of the bulk water contained at the core of the reverse micelle. "Our studies have revealed," they said, "that in the PFPE water-in-scCO$_2$ microemulsions, the 'bulk' water is sufficiently well formed to support solubility of ionic species, and spectroscopic data indicate that the species are in aqueous solution" (Clarke 1997, p. 6406).

Possible Applications

These reports are exciting news for industrial chemists. They suggest that supercritical carbon dioxide may find a wide variety of applications in industrial operations involving ionic, polar, and nonpolar compounds. Since carbon dioxide itself has no deleterious effects on the environment, its use as a solvent could solve a number of troubling problems that have developed with the use of organic solvents, such as the halogenated hydrocarbons.

References

Clarke, Matthew J., et al., "Water in Supercritical Carbon Dioxide Microemulsions: Spectroscopic Investigation of a New Environment for Aqueous Inorganic Chemistry," *Journal of the American Chemical Society*, 2 February 1997, 6399–406.

Cowen, R., "Carbon Dioxide Can Help Dissolve Proteins," *Science News*, 3 February 1996, 71.

Johnston, K. P., et al., "Water-in-Carbon Dioxide Microemulsions: An Environment for Hydrophiles Including Proteins," *Science*, 2 February 1996, 624–26.

Rouhi, Maureen, "Colorful Inorganic Chemistry Coaxed into Supercritical CO$_2$," *Chemical & Engineering News*, 11 August 1997, 40.

———, "Separations with Carbon Dioxide," *Chemical & Engineering News*, 29 September 1997, 6.

"Supercritical Carbon Dioxide: Uses as an Industrial Solvent," <http://www.ilsr.org/carbo/ps/factsh12.html> (20 December 1997).

Ward, Mark, "Greener Dry Cleaners," *Technology Review*, November–December 1997, 22.

Wu, Corinna, "A Green Clean," *Science News*, 16 August 1997, 108–09.

A Manganese Oxyiodide Cathode in Lithium Batteries

The demand for small, powerful, long-lasting batteries continues to grow. Any number of electronic devices, from portable radios and cellular telephones to laptop computers, operate on such batteries. Today, the battery of choice in such devices is a lithium-ion battery because the energy density (charge-to-mass ratio) of such batteries is higher than any other commercially available option.

One disadvantage of lithium batteries, however, is that they contain lithium cobaltite ($LiCoO_2$), a compound that is both expensive to make and toxic. These disadvantages have led to a number of efforts to develop other types of batteries with high energy density but lower cost and greater safety. In 1997, Jaekook Kim and Arumugam Manthiram at the Center for Materials Science and Engineering of the University of Texas at Austin reported on a new type of battery that meets these criteria.

The new battery has a cathode made of lithium, sodium, manganese, oxygen, and iodine with the putative formula of $Li_{1.5}Na_{0.5}MnO_{2.85}I_{0.12}$. An important feature of this material is that it can be made at room temperatures. This form of synthesis is significant because cathodes made at high temperatures (as most of them are) tend to have pockets of unreacted material that reduce the efficiency of the cathode to transmit an electrical current.

Kim and Manthiram prepared their new material by reacting anhydrous sodium permanganate ($NaMnO_4$) with lithium iodide (LiI) in acetonitrile. This approach ensured that the final product would be totally free from water. X-ray diffraction studies showed that the product of the reaction was amorphous, ensuring a particle size that would provide the highest possible energy density for its use as a cathode.

The life of the manganese oxyiodide battery appears to compare favorably with that of other lithium batteries. After more than 40 cycles of discharging and recharging, the researchers found that the battery was still producing current with an efficiency equal to that of a fresh product.

References

Kim, Jaekook, and Arumugam Manthiram, "A Manganese Oxyiodide Cathode for Rechargeable Lithium Batteries," *Nature*, 20 November 1997, 265–67.

"Manganese Oxyiodide Cathode Promises Cheaper, Safer Batteries," *Chemical & Engineering News*, 24 November 1997, 60.

The Structure of Flame Balls

Can you think of a *less* likely subject for chemical research than the structure of a flame? Humans have been using fire for thousands of years. Nearly every important industrial process and every form of transportation requires the combustion of some fuel. It would appear that flames are as well understood as anything in nature.

In fact, the behaviors of some kinds of flames are relatively well understood, and scientists have developed models of such flames that predict their actual behaviors with a fair degree of accuracy. But such is not the case for certain types of flames, such as those in which the ratio of fuel to air, or fuel to oxygen, is relatively low.

From a practical standpoint, however, such flames have many important everyday applications. For example, one way of meeting national and state regulations for reduced air pollution may be to use lean-burning hydrogen-fueled engines. Scientists currently know less about the behavior of such flames and the products they form than they would like.

A major problem in studying burning gases near the limits of their combustion ratios is the presence of gravity. Gravity is responsible for the fact that heated gases in a flame take on a tear-drop shape. Cool, heavy gases surrounding the flame flow downward and push hot, light gases within the flame upward, producing the classic shape of a candle or Bunsen-burner flame.

Flame Research in Low-Gravity Environments
As far back as 1944, the Russian scientist Yakov B. Zeld'ovich predicted that a flame produced in the absence of gravity would have a spherical shape. That is, a burning gas would produce a flame *ball* rather than the typical tear-drop shape with which we are familiar.

Until recently, there were no simple methods for testing Zeld'ovich's theory. One technique that has been used is known as a drop tower. A drop tower is a device in which an experiment can be enclosed in a cylindrical case and dropped from the top of a tower a few dozens of meters high. During this fall, the case briefly experiences a very low gravitational force known as a microgravity environment. Cameras mounted in the drop tower take up to 1,000 pictures per second, showing what happens to the material within the test chamber. Two widely used drop towers provide microgravity environments lasting 2.2 seconds in one tower and 5.18 seconds in the other.

A second technique for producing low-gravity environments is in aircraft that travel in a parabolic arc through the atmosphere. At the bottom of that trajectory, the aircraft experiences a gravitational environment about twice that on Earth, while at the top of the trajectory, the gravitational attraction drops to less than one percent of the Earth's gravitational pull.

Space Research
It comes as no surprise that the ideal setting for research on flame balls in microgravity environments is aboard spacecraft circling the Earth. In such settings, gravitational attraction drops to about 10^{-6} that on Earth. In addition, that state is maintained as long as the spacecraft remains in orbit.

Therefore, a group of experiments called SOFBALL (Structure of Flame Balls at Low Lewis-number), designed to study flame structure and properties in microgravity environments, was intended to be conducted on board the U.S. Spacelab contained in space shuttle flight STS-83, launched on 4 April 1997. Unfortunately, that flight was cut short because of a fuel-cell problem on the orbiter and only 2 of 15 experiments were actually carried out. The results of those experiments, however, were very exciting.

In the first experiment, a mixture of 4.9% hydrogen, 9.8% oxygen, and 85.3% carbon dioxide were ignited. As a result of the combustion, three flame balls were formed, each ranging in size from about 0.25 cm to 0.4 cm. (See figure 2.5.) The three balls remained essentially fixed in space and continued to burn for the duration of the experiment, a total of 500 seconds. In the second experiment, a mixture of 3.85% hydrogen and 96.15% air was ignited. Four flame balls formed in this case, lasting 150, 190, 270, and 320 seconds. Again, the flame balls remained essentially suspended in their original positions throughout the experiment.

The SOFBALL experiment was repeated on a later space shuttle flight, STS-94, launched on 1 July 1997. In this case, 12 of 13 gas mixtures ignited successfully, producing anywhere from 1 to 9 flame balls. Again, most flame balls lasted the 500 seconds set aside for the experiment and, when ignited a second time, burned for an additional 500 seconds. The properties of these flame balls were, in general, similar to those observed on STS-83.

Combustion in Flame Balls

The structure of a flame ball now appears to be relatively well understood. As combustion occurs, fuel gases and oxygen from the environment surrounding the flame ball move inward toward the surface of the ball. At the same time, heat and combustion products formed within the flame ball move outward. An equilibrium point is reached at which reactants and products are in balance with each other along the surface of the flame ball. This equilibrium accounts for the stability in size and shape of the flame ball.

Applications

The SOFBALL experiments have been motivated not only by scientists' curiosity about the structure and properties of flames, but also by some potentially important applications. Any fire that occurs in a spacecraft, for example, would have properties quite different from similar fires on the Earth's surface. The studies of flame balls in microgravity can help scientists understand, prevent, and deal with such potential disasters.

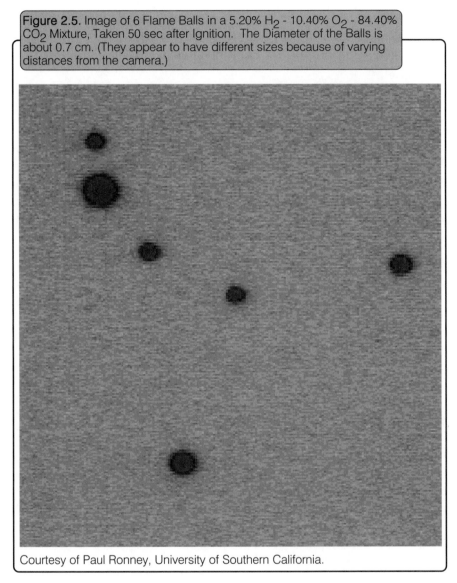

Figure 2.5. Image of 6 Flame Balls in a 5.20% H_2 - 10.40% O_2 - 84.40% CO_2 Mixture, Taken 50 sec after Ignition. The Diameter of the Balls is about 0.7 cm. (They appear to have different sizes because of varying distances from the camera.)

Courtesy of Paul Ronney, University of Southern California.

Also, scientists continue to investigate ways to make better use of fossil fuels so that smaller amounts of pollutants are released during their combustion. For example, the state of California has established very severe vehicle-emission standards to take effect in the year 2000. Many scientists believe that one of the best ways to meet those standards is to develop new types of engines in which fuels can be burned in a very low fuel-to-air mixture. Understanding the properties of flame balls is essential in developing such engines.

References

"MSAD Ground-Based Program," <http://ug.msad.hq.nasa.gov/cScienceProg/ground/groun.html> (26 September 1997).

Strauss, E. "Weak Flames Ignite Hope for Clean Engines," *Science News*, 26 July 1997, 54.

"Structure of Flame Balls at Low Lewis-number," <http://prometheus.usc.edu/sofball/> (26 September 1997).

"Structure of Flame Balls at Low Lewis-number (SOFBALL)," STS-83 Mission Results (initial), <http://cpi.usc.edu/sofball/sofball.html> (26 September 1997).

Amphiphilic Surfaces

In 1997, researchers at the University of Tokyo and TOTO, Ltd., under the direction of Akira Fujishima, reported on the discovery of a new kind of material to which water and oil bind equally well. This material shows great promise for use in antifogging and self-cleaning surfaces.

Scientists have known for some time that surfaces will coat easily with either water or oil, depending on which material is first applied to the surface. A water-binding surface is said to be *hydrophilic,* while an oil-binding surface is said to be *oleophilic.* However, a surface that binds equally well to both water and oil at the same time—an *amphiphilic* surface—had never been observed and proved to be an unexpected discovery.

The Research

The surface described here was prepared by coating a piece of glass with a thin layer of titanium dioxide (TiO_2). Both water and oil applied to the TiO_2 surface tended to bead up, indicating that neither coated the surface smoothly. The TiO_2 surface was then exposed to ultraviolet light and tested with both water and oil a second time. At this point, both water and oil showed contact angles of close to 0° with the surface. That is, they essentially wetted the surface consistently and smoothly.

The TiO_2 surface retained its ability to wet with both water and oil for a period of days, even when stored in the dark. Eventually, the surface returned to its pre-radiation appearance and beaded when wetted with either material. Exposure to radiation a second time, however, restored the surface's amphiphilic properties. The restoration of this property was apparently entirely dependent on its exposure to radiation and could be repeated any number of times.

Explanation

To understand this process, the research team examined the surface of the irradiated film with friction-force microscopy. They found clear and distinct domains, ranging in size from 30 to 80 nm, some bright in appearance and others dark. They identified the bright areas as hydrophilic regions and the dark regions as oleophilic regions. They concluded that the overall amphiphilic behavior of the surface on a macroscopic character was a function of the many discreet hydrophilic and oleophilic regions present at the nanoscale level.

Fujishima's team hypothesized that the conversion of the TiO_2 surface to an amphiphilic surface was caused by the conversion of Ti^{4+} ions to Ti^{3+} ions as a result of exposure to radiation. Ti^{3+} have been shown to lead to the dissociation of water molecules, permitting water to bind more closely to a surface. Over time and in the absence of radiation, Ti^{3+} reverts to Ti^{4+}, the surface's dissociative abilities are reduced, and the surface loses its ability to be wetted by water. Re-exposure to radiation once again converts Ti^{4+} to Ti^{3+} and restores the amphiphilic character of the surface.

Applications
Scientists are excited about the possible applications of this discovery. Amphiphilic surfaces remain transparent even when they are exposed to moisture. The moisture flattens out on the surface, rather than beading up and forming individual droplets. The surface, thus, remains completely transparent. Glass windows and mirrors with TiO_2 surfaces can be made resistant to fogging by this means.

Amphiphilic surfaces also tend to be self-cleaning. Fujishima's research group hung glass surfaces, some coated with TiO_2 and some not, outdoors for six months. They found the coated glass consistently cleaner than the uncoated glass. Oils and water droplets that normally leave residues on uncoated glass were washed away by the rain.

References
Wang, Rong, et al., "Light-Induced Amphiphilic Surfaces," *Nature*, 31 July 1997, 431–32.

Strauss, E., "Odd Companions Create Unusual Environment," *Science News*, 2 August 1997, 70.

COMBATING POLLUTION

Over the past half century, chemical research has produced a virtually endless supply of new products and processes that make human life safer, healthier, and more pleasant. But these accomplishments have also created a host of new social and technical problems, not the least of which is pollution. The manufacture, use, and disposal of so many synthetic products has been accompanied by the release of enormous amounts of long-lasting substances that may pose a threat to human health and/or safety or, at the least, create problems of disposal or destruction.

Small wonder, then, that one of the most active fields of chemical engineering today involves the search for methods by which the volume of pollutants released to the environment can be safely reduced. The three reports below illustrate the kinds of progress being made in this field.

Degradation of Halogenated Aromatic Hydrocarbons

Some of the most difficult pollutants to deal with involve aromatic compounds that contain one or more halogen atoms. Examples of such compounds include pesticides such as dichlorodiphenyltrichloroethane (DDT), 2,4-dichlorophenoxyacetic acid (2,4-D), chlordane, heptachlor, dieldrin, and aldrin. These compounds pose two major problems. First, they tend to be toxic to most forms of animal life. Second, they tend to be highly persistent in the environment.

Some kinds of microorganisms are able to degrade some kinds of halogenated hydrocarbons. However, such processes often take very long periods of time, and the volume of pollutants can easily exceed the capacity of natural populations of microorganisms to handle these pollutants.

In 1995, scientists at the Laboratorie de Chimie de Coordination du CNRS and the Elf-Aquitane Corporation, both in France, reported on a simple and inexpensive method for breaking down halogenated hydrocarbons. This method involves the use of two inexpensive and readily available compounds, hydrogen peroxide (H_2O_2) and a sulfonated metallophthalocyanine, a complex organic compound that contains an atom of either iron or manganese. The sulfonated metallophthalocyanine is represented as FePcS or MnPcS.

Reaction and Products

The compound chosen for study in this experiment was trichlorophenol (TCP). TCP can be regarded as an example of the kinds of halogenated aromatics released in many industrial operations. The benzene ring at its core does not degrade easily, and the chlorine atoms attached to the ring account for the compound's toxic effects.

The French research team used either H_2O_2 or $KHSO_5$ to oxidize TCP, in the presence of the catalyst FePcS. They measured the amount of TCP converted in 1- and 5-minute periods, using varying concentrations of $KHSO_5$ or H_2O_2, at pH levels of 2, 7, and 8.5. They also determined the structures of products formed in the oxidation process.

Results of the experiment indicated that both $KHSO_5$ and H_2O_2 can efficiently degrade TCP when FePcS is used as a catalyst. For example, TCP was converted with 100% efficiency in 1 minute or less by $KHSO_5$ in concentrations of 0.1 M (FePcS to substrate) and 1 M (MnPcS to substrate), at pH 2. A similar result was obtained in 5 minutes using H_2O_2 as an oxidizing agent and FePcS as a catalyst at pH 7. The latter result is perhaps somewhat more significant for this study since H_2O_2 is a common, inexpensive material with no environmental drawbacks.

Researchers also asked whether the products of the above reactions were environmentally safe. The degradation processes involved would be of little value, of course, if the products formed were not biodegradable or were as environmentally problematic as TCP itself.

Gas chromatography-mass spectrometry methods were used to identify the products of TCP oxidation. The four compounds identified as being formed by this process were malic, fumaric, chloromalic, and chlorofumaric acids. All of these compounds are environmentally less hazardous and far more easily biodegraded than is TCP.

The results of this research suggest that the H_2O_2–FePcS system can be used to degrade TCP to environmentally "friendlier" products quickly and at relatively low cost. Moreover, the system should be applicable to a very wide range of dangerous pollutants. The researchers themselves reported that they used TCP only as a model of the kind of compounds resistant to treatment. They proposed that "[t]his catalytic method should be considered as a credible chemical process for the oxidative mineralization of chlorinated phenols and might be useful for the industrial treatment of wastes containing not only polychlorophenols but also other recalcitrant pollutants, such as condensed aromatics" (Sorokin 1995, p. 1165).

References

Lipkin, R., "Taking Chlorine out of Tough Pollutants," *Science News*, 27 May 1995, 327.

Sorokin, Alexander, Jean-Louis Séris, and Bernard Meunier, "Efficient Oxidative Dechlorination and Aromatic Ring Cleavage of Chlorinated Phenols Catalyzed by Iron Sulfophthalocyanine," *Science*, 26 May 1995, 1163–65.

CFC Disposal

Chlorofluorocarbons (CFCs) have been implicated in the destruction of ozone in the stratosphere. Stratospheric ozone helps protect life on Earth from potentially harmful ultraviolet radiation in sunlight. Scientists, industrialists, and politicians worldwide are now generally convinced of the environmental risks posed by the manufacture and use of CFCs and have joined in a series of international agreements to eliminate the production and sale of CFCs.

The problem is that huge amounts of CFCs are stockpiled throughout the world. By one estimate, more than 100 million tons of one CFC,

Freon, is stored in the United States alone. Freon is banned for use in new cars and refrigeration systems, but is permitted for use in older systems. As a result, a thriving black market in CFCs has developed to move these compounds from stockpiles to consumers, often across international borders.

A technical problem exacerbating this situation is that, until recently, no inexpensive, simple method for the destruction of CFCs has been available. Two traditional methods have involved treating CFCs with molten sodium at high temperatures or burning the compounds. The first of these processes is relatively dangerous, however, and the second results in the formation of corrosive acids.

Sodium Vapor Reduction of CFCs

Another system for the destruction of CFCs, reported in 1997, involves a modification of earlier sodium-reduction techniques. Douglas P. Dufaux and Michael R. Zachariah at the National Institute of Standards and Technology (NIST) devised a system for reacting CFCs and sodium in the vapor phase at atmospheric pressure. The products of this reaction were environmentally benign halides of sodium and elemental carbon.

The NIST experiment was carried out in a reactor consisting of four concentric tubes carrying the reactants, cooling gases, and carrier gases. Sodium metal was heated to a temperature of 950 K and then introduced into the reaction chamber through one of the concentric tubes. Carbon tetrafluoride (CF_4) was introduced into the reaction chamber through a second tube. CF_4 was chosen as a model compound because it was known to be the most thermodynamically stable of all CFCs. Any destruction process that worked with CF_4 should, therefore, be effective with any CFC.

Sodium vapor and CF_4 were allowed to mix in the reaction chamber, which was maintained at a temperature of 1,100 K. In six repetitions of the experiment, the flow rates of CF_4 and sodium vapor were varied to determine the conditions under which decomposition occurred most efficiently. The least efficient of the arrangements studied still resulted in an 86% efficient conversion of CF_4, while the most efficient produced 100% conversion.

Results and Significance

The NIST researchers found that destruction of CF_4 with sodium vapor occurred with greater than 96% efficiency under 5 of the 6 experimental conditions tested. The products formed in the experiments consisted of sodium fluoride (NaF) and elemental carbon only. These products left the reaction chamber in the form of granular carbon particles covered with a thin layer of NaF. The particles ranged in size from about 200 to 500 nm.

Particles of this size can easily be separated by standard procedures from the gas in which they are carried.

The researchers suggested that the destructive technique described in this report could easily be adapted to industrial operations. They predict that the scale-up process should introduce no new technical problems. Furthermore, the reaction process can easily be converted to a closed-loop system in which sodium and carrier gases are continuously recycled, and the only net effect is the conversion of CFCs to elemental carbon and halide salts.

References
Dufaux, Douglas P., and Michael R. Zachariah, "Aerosol Mineralization of Chloro-fluorocarbons by Sodium Vapor Reduction," *Environmental Science and Technology*, August 1997, 2223–28.

"Sodium Vapor Destroys CFCs in Flame Process," *Chemical & Engineering News*, 11 August 1997, 35.

CHAPTER THREE
Social Issues in Chemistry

The development of the chemical sciences over the past two centuries has influenced human civilization in two fundamental ways. First, in contributing to the spread of a scientific philosophy of the world, it has provided humans with an alternative to much older world views in which events are perceived as being controlled by chance and unknown and unknowable random acts of capricious spirits, demons, gods, and goddesses. The scientific viewpoint has revealed to humans that the natural world can be understood in a rational way.

Second, from a more practical standpoint, chemistry has brought human civilization a staggering array of new materials that do not exist in the natural world: drugs and medicines, cosmetics, pesticides, fertilizers, building materials, fibers, dyes, and plastics, for example.

One could argue that the first of these changes is the more significant in that it has dramatically altered the way humans view nature and their place in the natural world. But it is probably the second change about which most of us are conscious on a day-to-day basis. One can hardly get through a day now without wearing, eating, or using a host of the products of chemical research and development. This chapter will deal with only the second of these developments—the production and use of new materials.

RECALLING THE HYDRA

In Greek mythology, the Hydra was a many-headed monster which, when one head was cut off, grew two in its place. The invention and manufacture of new materials has some similarities to a battle with the Hydra. Important chemical breakthroughs generally involve the discovery of a new material that will perform better than any existing comparable material . . . or that will do things no natural material can do. Such new materials have been used to solve a wide range of problems: from making "wrinkle-free" clothing and lighter, stronger aircraft bodies possible to finding cures for diseases. But the solution to each new problem is often accompanied by the introduction of other new problems that did not exist before the new material was discovered or invented. For example, the manufacture of the new material may produce harmful byproducts that result in air or water pollution. Or the use of the new material in products may create new waste-disposal issues for the products.

One might argue that many social, political, economic, and ethical issues facing the United States and much of the rest of the world today have some basis in scientific—including chemical—development. It is, therefore, unreasonable to do a thorough review in this chapter of *all* of the social issues raised by chemical research and development. Instead, the chapter focuses on a small number of especially interesting issues that have recently been of concern to the general public.

The following discussions are not intended to provide a complete and detailed description of the science involved in each issue. Such discussions are readily available in a number of standard sources. Instead, a minimal background is provided to remind readers of the science and technology involved in each case. Then, the major focus of the discussion shifts to recent developments in the field, with particular attention given to political, social, environmental, and economic considerations.

ATMOSPHERIC ISSUES

The 1960s and 1970s were an era of heightened human concerns for the environment. Possibly for the first time in human history, large numbers of people became aware of the dangers of pollution *and* were concerned enough about such dangers to demand governmental action. It was during this era that landmark legislation, such as the National Environmental Policy Act (NEPA) of 1969; the Resource Recovery Act (RRA) of 1970; the Resource Conservation and Recovery Act (RCRA) of 1976; the Clean Air Acts of 1963, 1965, 1970, and 1977; and the Comprehensive

Environmental Response, Compensation, and Liability (Superfund) Act (CERCLA) of 1980 were passed into law.

To a large extent, this new consciousness of environmental issues came about as a result of relatively local issues. For example, popular swimming beaches were closed because of polluted waters; people were advised not to leave their homes because of especially severe air pollution; and traditional fishing spots were declared off-limits because of mine runoffs.

The 1980s and 1990s saw increasing concerns about environmental issues at a new level of potential impact—issues that could affect the entire planet on a massive scale. Two of the most discussed of these issues were climate change (or *global warming*) and destruction of the ozone layer. The following two sections summarize recent developments with regard to these two issues.

Climate Change

Human life on Earth is possible at least partly because of the phenomenon known as the greenhouse effect. Sunlight striking the Earth's atmosphere warms the atmosphere by means of a process similar to that which occurs in a greenhouse. The Earth's "greenhouse story" is as follows.

Part of the sunlight that strikes the upper layers of the atmosphere is reflected back into space and plays no role in the warming effect. Another part is absorbed by gas molecules in the atmosphere, raising the atmosphere's temperature. A third part passes through the atmosphere and strikes the Earth's surface.

Sunlight striking the Earth's surface may be absorbed by land and water, or it may be reflected back into the atmosphere. Reflected sunlight may pass all they way through the atmosphere and back into space, or it may be captured by gas molecules in the atmosphere. In the latter case, this energy also contributes to an increase in the temperature of the atmosphere.

Various chemical species are more or less efficient at capturing energy reflected from the Earth's surface. One of the most efficient species at capturing this energy is carbon dioxide. One could reasonably argue that the more carbon dioxide in the atmosphere, the more solar energy reflected from the Earth is trapped, and the warmer the atmosphere becomes.

Although the most important, carbon dioxide is not the only gas capable of capturing solar energy reradiated from the Earth's surface. Other gases with the same property are classified along with carbon dioxide as *greenhouse gases*.

Issues

There is relatively little disagreement among scientists about the greenhouse scenario described thus far. The purported effects of human activities on the greenhouse effect and the ultimate consequences of those effects is the subject of greater debate.

Since 1958, scientists have been noticing a steady annual increase in the amount of carbon dioxide in the atmosphere. This trend is hardly surprising. Along with water vapor, carbon dioxide is the primary substance produced during the combustion of fossil fuels (coal, oil, and natural gas). As humans continue to burn fossil fuels at an accelerating rate (as they have for two centuries), the amount of carbon dioxide released to the atmosphere would be expected to increase proportionately. To many scientists, this increase in carbon dioxide concentration would also be expected to produce an increase in the planet's annual average temperature, a long-term global warming.

Other scientists and nonscientists disagree with this somewhat simplistic interpretation of increasing carbon dioxide levels. They argue that changes in cloud cover and in carbon sinks (such as tropical forests) have effects on global carbon dioxide patterns that we do not as yet fully understand.

The fundamental problem is that changes in the Earth's annual temperature observed so far are very small, well within those variations that take place as the result of natural phenomena over long periods of time. That is, it is still impossible to say with certainty whether the global warming that has been observed over the last two decades is a natural fluctuation or the result of human activities.

Furthermore, any possible effects of global warming are difficult to predict. One can extrapolate from existing data that an increase of $X°C$ will cause Y cubic kilometers of ice to melt with a consequent rise of Z centimeters in ocean level. One can then predict a variety of climate changes that might result from these changes. But these extrapolations are based on temperature changes of no more than a degree or so, over many decades, across the whole planet. Such predictions obviously and inherently involve a high degree of uncertainty.

In fact, some critics of the above theory say the degree of possible error in such extrapolations and predictions is much too high to allow for social, political, and economic decisions about any possible global warming.

That argument is crucial because global warming is not a purely scientific issue. It is ultimately a question as to what, if anything, humans should do about their current patterns of fossil-fuel combustion. If those who believe in global warming are correct, it would make sense to take actions soon to reduce fossil-fuel combustion. If their opponents are correct, it would be social, political, and economic nonsense to disrupt the world's economies to avoid a nonexistent problem.

Recent Developments

By the early 1990s, sufficient concern had developed about the problems of global warming that scientists and politicians around the world began to discuss the possibility of joint action to reduce carbon dioxide emissions. As a result of these discussions, the United Nations Conference on Environment and Development was scheduled for Rio de Janeiro in 1992. One hundred sixty nations were represented at this so-called Earth Summit. With only one exception—the United States—participants expressed a general agreement to cut back carbon dioxide emissions to 1990 levels by the year 2000. Special exceptions were provided for developing nations, who were given more time to reach the same goal and who were also promised an infusion of technology and financial help to aid them in reaching this goal.

Within a matter of years, however, it became clear that the voluntary approach to carbon dioxide emission reductions adopted at Rio had little or no chance of success. Carbon dioxide emissions continue to grow in nearly every industrialized nation. As a result, a new effort was begun to create a stronger, more binding agreement on carbon dioxide emissions.

An early step in that effort was the Berlin Mandate of 1995, in which the world's industrialized nations agreed to establish a system for limiting and cutting back on carbon dioxide and other greenhouse gas emissions. Signers of the Berlin Mandate also agreed to meet again in Kyoto, Japan, in December 1997, to work out the details by which reductions would be achieved. The Kyoto conference was officially designated as the Third Session on the Conference of the Parties to the UN Framework Convention on Climate Change.

At the planning sessions for the Kyoto meeting held in Bonn, Germany, chances for international agreement on global warming issues appeared slim. Participants disagreed widely on at least four basic issues. The first of these issues was how a "greenhouse gas" was to be defined. The United States and Canada argued that carbon dioxide, methane, nitrous oxide, hydrofluorocarbons, perfluorocarbons, and sulfur hexafluoride should all be defined as greenhouse gases. Other nations wanted to include only the first three of these compounds in that definition.

There was also disagreement over so-called emissions trading. Emissions trading refers to the policy whereby reductions in greenhouse gas emissions accomplished by one nation can be "given away" to another nation. For example, suppose that a firm in the United States sells a new technology for reducing greenhouse gas emissions to Kenya. Then Kenya may choose to give away any reduction in greenhouse gas emissions it accomplishes with that technology to the nation from which that technology was purchased, the United States.

So-called bubble practices are also a form of emissions trading. In a bubble system, some nations may increase greenhouse gas emissions while others decrease such emissions, as long as the overall net effect among all nations is a decrease in their total emissions. At the Bonn planning sessions, members of the European Union (EU) argued strongly for a bubble approach to greenhouse gas emissions limits in their region.

A third disagreement focused on the question of greenhouse gas emissions limitations for developing nations. The developing nations argued that industrialized nations currently produce three-quarters of all greenhouse gases and should, therefore, be responsible for controlling those emissions. Some industrialized nations, while acknowledging that fact, pointed out that developing nations are increasing their release of greenhouse gases at a rate much greater than that of industrialized nations and that by the year 2030, they will match the release of such gases by industrialized nations.

A fourth, and perhaps most important, difference of opinion at Bonn was the extent to which greenhouse gas emissions should be reduced. The EU arrived at the meetings recommending a 15% reduction below 1990 levels by the year 2010, the United States suggested cutting back to 1990 levels within the period 2008–12, and Australia suggested allowing an increase of 18% over 1990 levels. The most severe recommendations came from the Alliance of Small Island States (ASIS), whose very land-masses would presumably be most dramatically affected by a rise in ocean levels. The ASIS recommended a 20% cut in greenhouse gas emissions by the year 2005. Other nations arrived at Bonn with proposals ranging somewhere within the EU to ASIS range.

A key to the success of these negotiations was the U.S. position, which was reached by intense debate and discussion within the Clinton Administration. A task force appointed by President Clinton presented him with three options: stabilizing emissions at 1990 levels by 2010, by 2020, or by 2030. Clinton eventually decided to select a hybrid of these suggestions, choosing a target period of 2008–12 for stabilization at 1990 levels.

Strong opposition within the United States developed to the administration's position even before the Kyoto conference began. For example, the U.S. Senate unanimously adopted the Byrd-Hagel Resolution, which declared that the Senate would not ratify a global warming treaty that failed to include emission limits for developing nations. Also, dire warnings about the economic effects of placing *any* limitations on greenhouse gas emissions were issued by a number of companies and industry groups. For example, the Global Climate Coalition, an association of coal, oil, automobile, and chemical groups, warned that the U.S. position would lower the nation's production by three percent and result in the loss of more than three million jobs.

The Kyoto Conference

For the first 10 days at Kyoto, it appeared that disagreement over the 4 issues described above—and others of lesser magnitude—were going to be too great to allow consensus and the signing of an agreement. During the last 24 hours of the conference, however, powerful leadership by conference chair Raúl Estrada-Oyuela of Argentina brought agreement on all major issues, and a treaty was signed by 159 nations on 11 December 1997. The main elements of that treaty are as follows:

- Reduction of greenhouse gas emissions by 38 industrialized nations from 5 to 8% below 1990 levels by 2008–12
- Designation of six gases as greenhouse gases: carbon dioxide, methane, nitrous oxide, hydrofluorocarbons, perfluorocarbons, and sulfur hexafluoride
- Exclusion, for the time being, of developing nations from any greenhouse gas emissions limitations
- Provision for a variety of emissions trading among nations
- Creation of an approval period, from 16 March 1998 to 15 March 1999, during which nations may sign the treaty. The treaty will have taken effect when 55 nations that account for a total of 55% of the world's carbon dioxide emissions have ratified the treaty.

The fate of the Kyoto agreement is not yet entirely clear. For example, there appears to be *no* possibility of its being ratified in the United States because of the Byrd-Hagel Resolution. On the other hand, some major energy corporations are already beginning to make changes in their policies and practices that will lead to the goals established in Kyoto. British Petroleum (BP), for example, announced in May 1997 that it would increase by a factor of 10 its current $100 million investment in solar technologies and other efforts to reduce greenhouse gas emissions. In October 1997, Royal Dutch Shell announced a similar plan, which involves the investment of $500 million in alternative energy technologies.

References:

Hileman, Bette, "Global Climate Change," *Chemical & Engineering News*, 17 November 1997, 8–16.

_____, "Kyoto Climate Conference," *Chemical & Engineering News*, 22 December 1997, 20–22.

Mahlman, J. D., "Uncertainties in Projections of Human-Caused Climate Warming," *Science*, 21 November 1997, 1416–17.

Newton, David E. *Global Warming: A Reference Handbook*. Santa Barbara, CA: ABC-CLIO, 1993.

Ozone Depletion

The second major atmospheric problem of international concern is ozone depletion. Ozone is an allotrope of oxygen whose molecules consist of three atoms of oxygen (O_3) rather than just two, as found in the more common diatomic oxygen (O_2).

Ozone is formed in the Earth's upper atmosphere by series of reactions, the first of which occurs when solar energy causes the decomposition of diatomic oxygen:

$$O_2 \rightarrow O + O$$

The monatomic oxygen (O) produced then combines with a second molecule of diatomic oxygen to form ozone:

$$O + O_2 \rightarrow O_3$$

This reaction is sufficiently efficient to produce a layer in the stratosphere—the *ozone layer*—in which the concentration of ozone is significantly greater than it is elsewhere in the atmosphere. This ozone layer extends across most of the stratosphere, from an altitude of about 20 km to 50 km above the Earth's surface. Although this layer is regarded as being rich in ozone, the absolute amount of the gas is remarkably low. In fact, were it possible to transfer all of that ozone to the Earth's surface, it would form a layer no more than about three mm thick.

The ozone layer is very important to the survival of life on Earth because ozone molecules absorb a form of ultraviolet radiation (UV-B) that has been shown to have seriously harmful effects on both plants and animals. In humans, for example, UV-B radiation has been implicated in the development of skin cancers and cataracts and other forms of eye disorders. Cell and tissue damage in other animals and in plants has also been attributed to UV-B radiation.

In the mid-1980s, evidence began to accumulate that ozone concentrations in certain parts of the atmosphere were being substantially reduced. The region in which this effect was first observed was above the Antarctic. Scientists began to refer to an "ozone hole"—a vertical column above the Antarctic—where ozone concentrations plummeted far below their normal levels during the austral (southern-hemisphere) spring. Retrospective studies showed that this phenomenon had actually been occurring as far back as the early 1970s.

At first, scientists were unable to explain the occurrence of this ozone hole. Both physical (massive air movements) and chemical (reactions of ozone with other chemical species) causes were suggested. Eventually, experimental and observed evidence confirmed a chemical explanation. Specifically, it was shown that synthetic, chlorine-containing compounds produced and consumed on the Earth's surface were escaping into the stratosphere. At this altitude, these compounds broke down to release free chlorine atoms, which have the potential to react with and destroy ozone molecules:

$$Cl + O_3 \rightarrow ClO + O_2$$

$$ClO + O \rightarrow Cl + O_2$$

The first group of compounds implicated in this sequence of events was the chlorofluorocarbons (CFCs), a family of halogenated hydrocarbons that had become enormously popular for a variety of commercial and industrial applications.

Current Issues

Early discussions about ozone depletion mirrored those about global warming. Many questions were raised about the reliability of scientific data, about proposed explanations for ozone loss, about possible consequences of an ozone-depleted atmosphere, and about the economic costs of taking action to limit, reduce, or eliminate ozone loss. Remarkably, the direction of those discussions turned out to be very different from those about global warming. As annual reports showed continued and sometimes dramatic deterioration of the ozone concentration in the stratosphere, the world's scientists and politicians very quickly reached a consensus: Forceful action would be needed to reduce and eventually eliminate the release of CFCs to the atmosphere.

In 1987, a group of 23 nations signed the Montreal Protocol on Substances that Deplete the Ozone Layer. Eventually, well over 100 more nations signed the treaty, ensuring that it had, for all practical purposes, the unanimous support of the planet's nations.

According to the Montreal Protocol, the production and consumption of CFCs and other ozone-depleting chemicals was to be phased out according to a very specific timetable. The production of the most popular CFCs, for example, was supposed to be reduced by 20% by 1995 and by 50% by 2000.

The protocol had been in existence less than three years, however, when modifications were adopted. Evidence that ozone depletion was accelerating prompted the world's nations to set even more severe restrictions on the production of ozone-depleting chemicals in the London Amendments of 1990 and the Copenhagen Amendments of 1992. According to the latter agreement, for example, the production of CFCs was to be completely phased out by 1996.

Recent Developments

Efforts to solve the problem of ozone depletion have been remarkably successful. In fact, they have been widely praised as evidence that the nations of the world can work together to solve planet-wide problems.

By 1996, evidence for the success of this effort began to appear in the scientific literature. A paper in the 31 May 1996 issue of *Science* reported the first decline in the concentration of chlorine released by the breakdown of CFCs in the stratosphere. These results, however, did *not* mean that ozone depletion had ended. Indeed, studies reported in late 1996 showed that the ozone hole above the Antarctic was as large as it had ever been.

The problem is that CFCs can exist for a very long time in the atmosphere, often dozens or even hundreds of years. Even as their overall concentration declines, enough CFC (and other ozone-depleting) molecules remain in the stratosphere to attack ozone molecules. By some estimates, recovery of the ozone layer will not begin before 2005–10, and complete recovery may not be possible until the end of the twenty-first century.

In the meanwhile, some concerns about ozone depletion remain. In the first place, large quantities of CFCs and other ozone-depleting chemicals remain on the Earth's surface. For example, a plant legally operating in Monterrey, Mexico, continues to produce CFCs. Authorities are certain that some of these legally produced chemicals are illegally imported into and used in the United States. According to some experts, the continued availability of illegally obtained CFCs could extend the time at which ozone recovery is complete by many years.

Another impediment to a more rapid recovery of the ozone layer is the continued legal production and use of ozone-depleting substances other than CFCs. Methyl bromide, for example, is used primarily as a fumigant on domestically produced crops such as grapes, raisins, cherries, nuts, and on other fruits and vegetables. It is most commonly injected into the soil before a crop is planted, but it is also used as a spray on products after they have been harvested. Concerns about the use of methyl bromide on agricultural products is based on the fact that an estimated 80–95% of the material eventually escapes into the atmosphere, where it can cause health effects in humans and environmental damage to the ozone layer.

Under provisions of the Montreal Protocol and its amendments, the production and use of methyl bromide are to be phased out by the year 2005, although some exemptions are to be permitted. For example, the compound is not banned in developing countries until the year 2015. In the United States, however, the Clean Air Act requires a complete ban of methyl bromide in the United States by the year 2001, with *no* exceptions permitted.

In mid-1996, the issue of methyl bromide production and use in the United States heated up. American farmers complained that no adequate substitute for methyl bromide yet exists, and none is likely to become available before 2001. U.S. farmers will, therefore, be at a serious disadvantage compared with farmers in other parts of the world, where the compound will still be allowed, especially with respect to farmers in developing countries.

Environmentalists point out, however, that bromine is about 50 times as efficient at destroying ozone as is chlorine. The campaign that has thus far been so successful in reversing the loss of stratospheric ozone should not, they argue, be hindered by allowing the continued production and use of methyl bromide.

References:

Brown, George E., Jr. "Environmental Science Under Siege in the U.S. Congress," *Environment*, March 1997, 12–20+. Also see comments on this article in *Environment*, May 1997, 3–5+.

Karliner, Joshua, "The Barons of Bromide: The Corporate Forces Behind Toxic Poisoning and Ozone Depletion," *The Ecologist*, May/June 1997, 89–90.

"Methyl Bromide Information," <http://www.igc.qpc.org/panna/MBPage.html> (14 September 1997.).

Newton, David E. *The Ozone Dilemma: A Reference Handbook.* Santa Barbara, CA: ABC-CLIO, 1995.

Clean Air

The 1960s might well be called the Decade of Environmentalism. The rise of environmental consciousness began with concerns about air and water pollution, and it was in these two areas that some of the earliest environmental legislation was passed. The Clean Air Act of 1963 was based on an even earlier model, the Air Pollution Control Act of 1955, and was the basis of later laws, such as the Air Quality Act of 1967 and its amendments in 1970, 1977, and 1990.

This family of laws is complicated because they deal with a variety of air pollution problems, including acid precipitation, urban smog, ozone depletion, and hazardous air pollutants. In terms of health effects on humans, however, the cornerstone of the acts is a set of standards called the National Ambient Air Quality Standards (NAAQS). These standards, which are designed to protect human health, establish concentration limits for six air pollutants—ozone, nitrogen oxides, sulfur dioxide, carbon monoxide, lead, and particulate matter.

"Protecting human health" does *not* mean completely eliminating any possible threat to human health from these pollutants. Instead, it means reducing the risk of disease as much as reasonably possible for the most sensitive of American citizens: children, the elderly, and those with respiratory disorders such as emphysema and asthma. According to some estimates, the individuals who make up these groups constitute about a third of the U.S. population.

The Clean Air Act is generally regarded as one of the most successful pieces of environmental legislation ever passed in the United States. In the 25 years following passage of the 1970 act, the concentration of air pollutants has dropped anywhere from 28% (for carbon monoxide) to 98% (for lead). Only in the case of nitrogen oxides was there an increase (6%) during the same period.

Current Issues
The Clean Air Act requires the U.S. Environmental Protection Agency (EPA) to review the NAAQS every five years and to make revisions as it deems necessary, based on the best scientific information available at the time. In November 1996, the EPA announced its most recent plans to revise the NAAQS. This process of revision is a lengthy one, requiring the publication of the Agency's intentions for public review, followed by the publication of final rule-making statements.

The EPA's initial statement outlined plans to change standards for two pollutants, ozone and particulates. The ozone standards apply to ozone found in the lower troposphere, near the Earth's surface. In contrast to stratospheric ozone, which serves a useful function by protecting human health and the environment, tropospheric ozone is harmful to both human health and other organisms in the environment.

The EPA proposed reducing permissible levels of ozone from 0.12 ppm (parts per million) measured over a one-hour period to 0.08 ppm measured over an eight-hour period. It also proposed changing the particulate-matter standard by adding a new category of ultrafine particles with diameters of less than 2.5 μm. The standards for such particles were set at 15 μg per cubic meter annually and 65 μg per cubic meter in a 24-hour period.

These proposals were met by strong opposition from both businesses and legislators. In particular, oil, chemical, mining, and paper companies complained that the cost of meeting the new EPA standards would be unacceptably high. In July 1997, the American Trucking Association, the U.S. Chamber of Commerce, and the National Coalition of Petroleum Retailers joined together to file suit against the EPA, asking the court to overturn the Agency's new standards.

EPA supporters pointed out that industry has traditionally expressed concerns about its costs for air-pollution control and that it has usually overestimated these costs substantially. In the battle over the Clean Air Act of 1990, for example, industry predicted that provisions of that law would cost them between $51 billion and $91 billion per year. According to EPA estimates, however, the actual costs to industry of implementing that act will be about $22 billion, once it has been fully implemented in 2005. These actual costs are 57–76% less than industry's original projections.

A second complaint voiced by both industry and some legislators was that the new EPA NAAQS are not based on sound scientific information. Senator James M. Inhofe (R-OK) claimed that the standards were "arbitrary, unfair, and unwarranted [and] they are based less on scientific and public health than they are on politics" (Raber 1997, p. 26). A number of Inhofe's colleagues in the House of Representatives agreed with this position and more than 130 cosponsors were obtained for the bill H.R. 1984, which would have delayed implementation of the new EPA standards until the next five-year cycle ends. The bill also authorized $75 million annually for research on the health effects of ultrafine particulate matter.

A third issue relating to the EPA's new NAAQS involved the 1996 Small Business Regulatory Enforcement Fairness Act (SBREFA). This act requires that federal agencies conduct studies of potential impacts of any new rule-making on small businesses as well as on state, local, and tribal governmental bodies. EPA critics argue that the Agency bypassed this law in developing its 1996 NAAQS changes. EPA representatives explained that they were aware of SBREFA requirements, but that NAAQS is a set of national standards that, the courts have ruled, do not fall under the provisions of the Small Business Act.

References:

"Clearing the Air: EPA's Tightening of Air Pollution Standards," *Consumer Reports*, August 1997, 36–38.

"Clinton Accepts New Clean Air Proposals," *Science News*, 5 July 1997, 6.

Crandall, Robert W., "EPA's Proposed Air Quality Standards: Clean Air Sense," *Brookings Review*, Summer 1997, 40–47.

"Overview of the Clean Air Act," <http://www.webcom.com/staber/caa.html> (30 December 1997).

Raber, Linda R., "Foes Take Aim at New Air Standards," *Chemical & Engineering News*, 11 August 1997, 26–27.

"Tiny Particles, Big Dilemma," *Business Week*, 4 August 1997, 82–83.

PROTECTING THE NATION'S FOOD SUPPLIES

One of the most direct impacts of chemical research on the lives of ordinary Americans is in the foods we eat. In the not-so-distant past, almost all the foods that Americans ate were grown at home or obtained from a farm less than a hundred miles away. Most of those foods were "natural," or close to it; that is, few chemicals had been added to them artificially during growing and processing.

Today, the story is far different. Virtually any kind of food produced anywhere in the world is available to Americans at any time of the year. This transformation in modern nutrition has been made possible by the discovery and invention of chemicals that preserve foods over long periods of time, that kill pests that destroy foods, and that enhance the properties of foods, such as color, odor, taste, and nutritional value.

For more than four decades, the potential health risks of pesticides and food additives have been well known. Some chemicals developed to protect crops or alter the properties of foods have later been found to be carcinogenic. As early as 1906, the U.S. Congress passed the Pure Food and Drug Act. That act was amended in 1938, but the federal government did not concern itself with chemicals added to foods until a 1958 revision of that act.

In 1972, Congress also passed the Federal Insecticide, Fungicide, and Rodenticide Control Act (FIFRCA). The primary goals of this act were to protect farm workers, who use pesticides, and the general public, whose food supplies commonly contain some residues of pesticides used in food production. These two laws have turned out to be anything but perfect, but they have functioned surprisingly well in protecting the nation's food supplies.

One of the most controversial parts of the nation's food protection policies has been an amendment added to the Pure Food, Drug, and Cosmetic Act of 1958 by Representative James J. Delaney (D-NY). The so-called Delaney Clause prohibited the use of any food additive if that additive had been shown to cause cancer in experimental animals at *any* concentration, no matter how low. The Delaney Clause established a "no-risk" or "zero-tolerance" policy in that it did not provide for any exceptions whatsoever.

For example, suppose that chemical A had been found to produce cancer in mice when it was fed to them in a concentration a thousand times greater than that which a human would receive in a whole lifetime. According to the Delaney Clause, that chemical could not be used as a food additive *at all*, even in concentrations far less than scientists today believe could possibly cause cancer.

In addition, the Delaney Clause applied to all food additives, but to virtually no pesticides. It regulated only those pesticides whose concentration *increased* during food processing. Since the amount of pesticide remaining in foods nearly always remains constant or decreases, the Delaney Clause applied to only the smallest fraction of pesticides.

Over the years, dissatisfaction with the Delaney Clause grew. For one thing, technology passed the standard by. Chemists have developed the ability to detect food additives at a level far more precise than had been possible in 1958. Additives present in nanogram quantities in food had to be banned in 1990 even though a chemist in 1960 would never even have known they were present. In unspoken recognition of this fact, the Food and Drug Administration and the Environmental Protection Agency slowly began to ignore the Delaney Clause in practice. Then, in the early 1980s, consumer advocates filed suit against this practice, and the Supreme Court eventually ruled that the Delaney Clause was still the law of the land and had to be enforced.

Concern about the Delaney Clause also grew because processed and unprocessed foods were being regulated differently. While the Delaney Clause was the standard for processed foods, unprocessed foods were judged by a risk/benefit standard. That is, chemicals with known potential for damage to human health were allowed in some foods provided that farmers could show severe economic hardships as a result of banning those chemicals.

The Food Quality Protection Act of 1996

In an attempt to deal with the issues surrounding food safety, H.R. 1627 was introduced into the 104th Congress. The bill quickly gained 240 cosponsors and seemed likely to pass the House easily. Some observers considered it to be an "industry bill," whose major effects would have been to dramatically decrease the regulation of chemical residues on foods.

This perceived bias galvanized opponents of H.R. 1627. In general, while the opponents acknowledged the need for reform of the nation's food additive and pesticide residues laws, they tended to have a somewhat different view as to what the final legislation should look like. Over a period of months, the two sides argued over the bill. Eventually, Congress rewrote large parts of the bill and then, in July 1996, passed H.R. 1627 in its revised form. The bill received unanimous approval from both houses of Congress.

The new Food Quality Protection Act of 1996, as signed by President Clinton on 3 August 1996, contains two major provisions. First, the same health standards are to be applied to all foods, whether raw or processed. A relatively small number of exceptions is provided for farmers producing certain foods. But foods produced under these special conditions must be so labeled in the marketplace.

Second, the Delaney Clause was replaced by a more general "safe-level" rule for chemicals. The "safe level" for any given additive or pesticide was to be determined by a method devised by the EPA. But that method had to take into account the greater sensitivity of infants and children to exposures and had to guarantee their safety under the standards adopted.

References:

Hattis, Dale, "Drawing the Line: Quantitative Criteria for Risk Management," *Environment*, July/August 1996, 10–15+.

Meyerhoff, A., and M. W. Pariza, "Should Pesticides that Cause Cancer in Animals Be Banned from Our Foods?" *Health* (San Francisco), March/April 1996, 36.

"The Nation's New Pesticide Law," <http://www.ewg.org/pub/home/reports/newpestlaw/newpestlaw.html> (26 September 1997).

"New Laws Rewrite Rules on Pesticides," *Science News*, 7 September 1996, 159.

S. 1166—Food Quality Protection Act, Hearing before the Committee on Agriculture, Nutrition, and Forestry of the United States Senate, 104th Congress, 2nd Session, 12 June 1996.

CHEMICAL WEAPONS CONVENTION

Humans have developed and used virtually every conceivable weapon in doing battle with each other, from sticks and stones to bows and arrows to almost every imaginable form of guns and bombs. Still, a few weapons that have been known to be powerful methods of mass destruction have been withheld by almost all nations under almost all circumstances. These weapons include nuclear devices and chemical and biological agents.

This reluctance to use the most horrendous weapons of all has not, of course, been universal or total. The United States, for example, is the only nation in history to have employed nuclear weapons in battle, and then only twice. The use of biological weapons has been extremely rare, although not unheard of. Perhaps the earliest recorded use of such weapons was 1346. A Mongol army laying siege to the Crimean city of Kaffia gathered together corpses of its own people who had died of the plague and catapulted them over the walls of the city. The trick was successful as the citizens of the city rapidly evacuated their homes and surrendered to the Mongols. In a more recent example of biological warfare, Japanese troops spread fleas infected with plague microorganisms over 11 cities in China at the beginning of World War II.

Still, the general horror over the use of nuclear, biological, and chemical weapons has historically been so great that the vast majority of nations have declined to use them in conflict even when those weapons were available and the obvious goal of battle was to kill as many of the enemy as quickly and efficiently as possible.

Chemical weapons have been defined to include any chemical which, through its chemical effects on living processes, can cause death, temporary loss of performance, or permanent injury to people and animals. Such materials can be classified into a half dozen major categories: nerve agents, mustard gases, tear gases, arsines, hydrogen cyanide, and psychotomimetic agents. Overall, some 70 different chemicals have been classified as actual or potential chemical weapons.

As with nuclear and biological weapons, the use of chemical weapons is rare, but not unheard of. As examples, Germany used chlorine gas against the French in World War I, Italians used chemical weapons in Ethiopia in 1937, and Egypt used chemical weapons in battles against Yemen in the 1960s. Today, one of the nations most generally suspected of making, storing, and using chemical weapons is Iraq. The Iraqis are known to have used hydrogen cyanide and mustard gas—and, possibly, other chemical weapons as well—against Iran in its 1980–88 war, against its own Kurdish citizens in 1988, and perhaps against U.S. and allied troops during the Gulf War.

An additional issue raised by the availability of chemical weapons is domestic attacks. In March 1995, the Japanese religious cult Aum Shinrikyo released the nerve gas sarin in a crowded Tokyo train station, killing a dozen people and injuring 5,000 more.

The Chemical Weapons Convention

Attempts to ban chemical and biological weapons (CBW) go back to the early 1920s. Horrified by Germany's use of chlorine gas during World War I, the nations of the world, through the League of Nations, adopted the Geneva Protocol, which outlined mechanisms for banning the production, storage, and use of chemical and biological weapons. Although numerous efforts were made, the United States never signed that treaty, and the Protocol failed to have any significant impact on CBW research and development.

Sufficient concern about the dangers of CBW remained, however, to keep the issue before statesmen around the world. This concern eventually led to the adoption in 1972 of the Biological Weapons Convention, an agreement signed by 87 nations, which banned the production, testing, and storage of biological weapons. The United States was a signatory to the treaty even though President Richard Nixon had announced three years earlier that the United States would discontinue further research on CBW.

The success in achieving a Biological Weapons Convention encouraged scientists and some politicians to seek a similar agreement on chemical weapons. Negotiations for such a treaty began in the early 1980s, largely under the leadership of U.S. representatives. Those negotiations continued through the administrations of Ronald Reagan, George Bush, and Bill Clinton, and a draft treaty was finally completed in 1993. Eventually, 162 nations signed this treaty, with a few important exceptions, including Iraq, Libya, North Korea, and Syria.

The Chemical Weapons Convention (CWC) was initially submitted to the U.S. Senate in November 1993 for ratification. The Senate failed to act immediately and, in fact, had totally ignored the CWC as of mid-1996. At that point, the Senate was virtually forced to take action since the treaty was scheduled to go into effect on 29 April 1997, regardless of whether the United States had ratified it.

For some time, the fate of the treaty was very much in doubt. The leading opponent of ratification was Senator Jesse Helms (R-NC) who raised a number of objections to the treaty. One was that it would place insufferable burdens on U.S. businesses. He argued that few small businesses "can afford to employ the legion of lawyers [needed] to satisfy the new reporting requirements" (Cooper 1997, p. A9) of the treaty. He warned that businesses as diverse as dry cleaning establishments, breweries, and the Mary Kay Cosmetics company would be damaged by provisions of the treaty.

Opponents of the treaty were also convinced that the United States' ability to develop deterrents to chemical weapons would be eliminated by the treaty since research using such materials would be banned. They also worried about the exchange of technical information among signatory nations that was required by the CWC, claiming that such exchanges might give other nations technical information about chemical weapons development that they could not obtain on their own.

Also, critics pointed out, some of the nations most likely to use chemical weapons—Iraq and Libya, for example—had not signed the treaty and would not be covered by its terms. Finally, opponents of the CWC simply did not believe that it was enforceable. As Helms repeatedly said, "It just will not work. This treaty will not work. If anything, it will make the problem worse" (Rankin 1997, p. A16).

Proponents of the treaty offered responses to this array of arguments and presented their own reasons for supporting the CWC. First, they pointed out that the purported threat to small businesses was a "red herring." Only businesses producing more than 30 metric tons of a banned chemical annually were required to file reports, and inspections were to be scheduled only for businesses producing more than 200 metric tons of such chemicals annually. A representative of the American Chemical Society testified that his group saw "little or no impact on small businesses" (Cooper 1997, p. A9) from the CWC.

Supporters of the treaty also noted that the United States was already destroying its stockpile of chemical weapons, so the treaty created no demands in this regard that were not already being met. They also said that the United States' enormous stockpile of conventional and nuclear weapons guaranteed the nation's position as the world's greatest military power.

Finally, supporters of CWC reminded the Senate that the CWC would go into effect in April 1997, in any case, so that the United States might very well be left on the outside looking in on the chemical weapons verification process. The obvious alternative was to ratify the treaty and become part of that process. In addition, supporters point out, if the United States did *not* ratify the treaty, the U.S. chemical manufacturers would be subject to export restrictions that would not be applied to treaty members.

In the end, the arguments of the supporters prevailed and the U.S. Senate voted 74 to 26 to ratify the Chemical Weapons Convention on 24 April 1997.

References:

"Chemical Weapons: Controlling International Production and Use," *Congressional Digest*, June/July 1997.

Cooper, Mary H., "The Fight to Ban Chemical Weapons," *The Oregonian*, 13 February 1997, A9.

Rankin, Robert A., "Chemical Weapons: The Treaty Test," *The Oregonian*, 24 April 1997, A16.

Towell, Pat, "Chemical Weapons Ban Approved in Burst of Compromise," *Congressional Quarterly Weekly Report*, 26 April 1997, 973–75.

CHAPTER FOUR
Biographical Sketches

W ho gets credit for important discoveries, inventions, and other breakthroughs in chemical research? Not so long ago, that question might have been relatively easy to answer. Scientific papers reporting on such breakthroughs were often signed by a single individual or, at most, by two or three scientists. Under those circumstances, it was comparatively easy to assign credit for important discoveries and inventions.

But times have changed. Today, most research is conducted by teams of scientists. Such teams are usually headed by a senior scientist with considerable experience in the field under investigation. The team may also include one or more postdoctoral students who have completed their Ph.D. research but are spending a year or two improving their skills and knowledge in some specific field of chemistry. The research team may include one or more graduate students also. In many cases, the work done by those graduate students will satisfy the research requirements for their own Ph.D. degrees. Finally, research teams occasionally include undergraduate students and technicians with special skills in the field under investigation.

It is also common for research teams from two or more institutions to work together on a single project. For example, a university research team may have special skills in the theoretical and conceptual aspects of a topic, while an industrial research team may have the technical skills and imagination to carry out relevant experiments.

Under these circumstances, it is often difficult to assign credit for an important breakthrough to a single individual. For example, the research report describing the discovery of the properties of element 106 listed 18 authors. When one *does* want to assign credit for a discovery or invention, that credit often goes to the leader of a research team or the team's senior scientist.

This chapter provides brief biographical sketches of some of the most important figures in chemical research over recent years. The sketches are provided with the understanding that many people contributed to the success of any one individual's work described here.

These sketches have the additional purpose of illustrating the wide variety of individuals with very diverse backgrounds now involved in chemical research. In the preparation of this book, an effort was made to obtain biographical information for at least one member of every research team whose work is presented in chapters 1 and 2. In some cases, that information was not available, but the majority of research teams mentioned in those chapters are represented by at least one person in this chapter.

Nancy C. Andrews

In 1997, Andrews led a research team that identified a gene that may be responsible for the production of an iron transport protein. Finding this gene and protein has significant potential benefits since hundreds of millions of people worldwide have medical problems associated either with the over-absorption and -utilization or with the under-absorption and -utilization of iron in their bodies.

Andrews earned her B.S. and M.S. in molecular biophysics and biochemistry at Yale University simultaneously in 1980. She then attended the Massachusetts Institute of Technology, from which she received her Ph.D. in biology in 1985. Two years later, she was also awarded her M.D. from the Harvard Medical School. Her postdoctoral training included stints at the Children's Hospital in Boston and the Harvard Medical School from 1987 to 1989 and at Dana-Farber Cancer Institute.

In 1991, Andrews accepted appointments as Instructor in Pediatrics at the Harvard Medical School and as Assistant in Medicine at the Children's Hospital and Pediatric Oncology of the Dana-Farber Cancer Institute. Two years later, she was promoted to Assistant Professor at Harvard and was appointed Assistant Investigator at the Howard Hughes Medical Institute. Since 1996, Andrews has also served as Associate Faculty Director of the M.D.-Ph.D. Program at the Harvard Medical School.

Peter Armbruster

Armbruster has been involved in the discoveries of most of the transfermium elements found over the last two decades. His primary field of interest has been heavy-ion physics, particularly the synthesis and production of new elements and isotopes and the development of reactions by which these materials can be made.

Armbruster studied physics at the Technical Universities of Stuttgart and Münich (1952–1957) and earned his Ph.D. at the latter institution in 1961. He worked at the Research Center of Jülich (1964–1970) and then accepted a position at the Gesellschaft für Schwerionenforschung (GSI) in Darmstadt. He is now a senior scientist at GSI.

In addition to his affiliation with GSI, Armbruster has served as Research Director of the European Institute Laue-Langevin at Grenoble, France (1989–92), and has served on the faculty at the University of Cologne (1968) and the Technical University of Darmstadt (1984–to date).

David Baker

In 1997, Baker led a research team at the University of Washington that determined the simplest possible protein that can form a three-dimensional beta-sheet structure and carry out essential biochemical functions. His team discovered that a functioning polypeptide containing just five different amino acids could be produced that fulfilled these requirements.

Baker earned his B.A. from Harvard University in 1984 and his Ph.D. from the University of California at Berkeley in 1989. In 1993, after completing a three-year postdoctoral program at the University of California at San Francisco, Baker accepted his current appointment as Assistant Professor in the Department of Chemistry at the University of Washington, in Seattle.

Paul F. Barbara

In 1997, Barbara and his colleagues reported on a phenomenon that could be described as "blinking molecules." They described the discovery of a substance that consists of a copolymer of poly(p-phenylene vinylene) and poly(p-pyridylene vinylene). When molecules of this substance were irradiated, all chromophore groups turned on and off almost simultaneously, producing a "blinking" effect.

Barbara earned his B.A. from Hofstra College in 1974, and his Ph.D. in chemistry from Brown University in 1978. He then did his postdoctoral work at Bell Laboratories from 1978 to 1980.

In 1980, Barbara joined the faculty at the University of Minnesota, where he has served as Assistant Professor (1980–86), Associate Professor (1986–90), Professor (1990–95), and 3M-Alumni Distinguished Professor (1995–to date). In addition to his academic work, Barbara has been a consultant to the 3M and Honeywell Corporations.

Paul D. Boyer

Boyer was awarded a share of the 1997 Nobel Prize in Chemistry with John E. Walker and Jens C. Skou. He was honored for his theoretical speculations about the way in which the enzyme ATP synthase operates in mitochondria. He suggested that a flow of hydrogen ions causes the cylindrically shaped enzyme to spin, in much the way that water forces a water wheel to spin on its axis. The spinning of the ATP synthase made it possible for the components of ATP to be brought together in such a way as to permit synthesis of the essential molecule.

Boyer received his B.S. from Brigham Young University in 1939 and his M.S. and Ph.D. in biochemistry from the University of Wisconsin in 1941 and 1943, respectively. His first teaching position was as an instructor in chemistry at Stanford University (1943–1945). He then served successively as Assistant Professor, Associate Professor, and Professor of Biochemistry at the University of Minnesota (1945–1956). He later returned to California to serve as Professor of Chemistry and Director of the Molecular Biology Institute at the University of California at Los Angeles.

Karen J. Brewer

Brewer's special research interests focus on the development of very large ("supramolecular") multimetallic complexes and the study of the spectroscopic and electrochemical properties of these compounds. She has produced more than 40 publications in this area over the past 4 years.

Brewer earned her B. S. in chemistry in 1983 from Wofford College in Spartanburg, South Carolina, and her Ph.D. in 1987 from Clemson University. She then worked for a year as a postdoctoral fellow in the laboratory of Melvin Calvin at the Lawrence Berkeley Laboratory of the University of California at Berkeley.

After leaving Berkeley, Brewer served as Assistant Professor of Chemistry at Washington State University, in Pullman, for four years. She then moved to the Department of Chemistry at the Virginia Polytechnic University and the State University at Blacksburg, where she has served as Assistant Professor (1992–95) and Associate Professor (1995–).

David W. Christianson

In 1997, Christianson reported the discovery of the crystal structure of the carboxypeptidase A enzyme. This paper was his latest of more than a dozen papers on this topic.

Christianson has served as Professor of Chemistry at the University of Pennsylvania since 1996. He received his A.B. in chemistry, *magna cum laude*, from Harvard College in 1983, and his A.M. and Ph.D. in chemistry from Harvard University in 1985 and 1987, respectively.

In 1988, he was appointed to his first teaching position at the University of Pennsylvania, where he served as Assistant Professor (1988–93) and Associate Professor of Chemistry (1993–96), before being appointed to his current position at the university. He has also served as an adjunct professor at the Wistar Institute since 1991, and as a consultant to a number of drug and chemical companies, including SmithKline Beecham, Alcon Laboratories, Lexin Pharmaceutical Corporation, Syntex, and DNX, Inc.

Paul Ching-Wu Chu

Chu is one of the world's leading authorities on high-temperature superconductivity. Research teams under his direction have, for more than a decade, developed a variety of materials that act as superconductors at increasingly high temperatures. In 1997, Chu coauthored a report on a new superconducting material that contained no toxic elements. This announcement represented an important step forward in making superconducting materials more readily available for a variety of commercial applications.

Chu received his B.S. from Cheng-Kung University in Taiwan in 1962. He then left Taiwan and enrolled at Fordham University in New York City, from which he received his M.S. in 1965. In 1968, he earned his Ph.D. in solid-state physics from the University of California at San Diego.

After receiving his doctoral degree, Chu worked for two years as a member of the technical staff at Bell Laboratories in Murray Hill, New Jersey. In 1970, he accepted an appointment as Assistant Professor in the Department of Physics at Cleveland State University, in Cleveland, Ohio, where he was promoted to Associate Professor in 1973, and to Professor in 1975. In 1979, he moved to the University of Houston, where he currently holds posts as Professor of Physics, Director of the Texas Center for Superconductivity, and the T.L.L. Temple Chair of Science. In addition to his academic appointments, Chu has held a number of important research positions, including Director of the Magnetic Information Research Laboratory at Houston (1984–88), Director of the Space Vacuum Epitaxy Center sponsored by Houston and NASA (1986–88), Director of the Solid State Physics Programs of the National Science Foundation (1986–87), and Director of the National Science Foundation Materials Research Science and Engineering Center on Advanced Materials at the University of Houston (1996–97).

Robert H. Crabtree

In 1997, a research team under Crabtree's direction reported the discovery of a new kind of hydrogen bond—a "dihydrogen bond"—that appears to form with certain boron-hydrogen and transition-metal-hydrogen.

Crabtree currently serves as Professor of Chemistry at Yale University. He earned his B.A. at Oxford in 1970 and his Ph.D. at Sussex University in 1973. He soon left the U.K. to carry out his postdoctoral research at Centre National de la Recherche Scientifique (CNRS) in France (1973–1977). He then moved to the United States, where he accepted a position at Yale.

Paul Crutzen

Crutzen shared the Nobel Prize for Chemistry in 1995 with Mario Molina and F. Sherwood Rowland for their research on chemical reactions that have led to depletion of the ozone layer in the stratosphere. This research showed that certain chemicals produced synthetically on the Earth have a tendency to break down to form species that react with and destroy ozone molecules. These chemicals—the chlorofluorocarbons (CFCs)—were discovered in the late 1920s and rapidly became enormously popular for a variety of commercial and industrial uses, including aerosols, air conditioning and refrigeration systems, and cleaning agents. For many years, scientists had little concern about any possible environmental threat from the CFCs because, at sea level, they are very stable. It came as something of a surprise, then, to discover that they became unstable when exposed to solar radiation in the upper atmosphere, breaking apart and forming ozone-destroying species.

Crutzen studied civil engineering in his native Amsterdam from 1951 to 1954. He then took a job with the Bridge Construction Bureau of the City of Amsterdam (1954–1958) and later served in the Dutch Army (1956–1958). In 1958, he took a job in the House Construction Bureau in Gävle, Sweden, before continuing his academic studies. Crutzen earned his M.Sc. in 1963 and his Ph.D. and D.Sc. in meteorology from the University of Stockholm in 1968 and 1973, respectively.

In 1974, Crutzen traveled to the United States, where he took a job as a research scientist in the Upper Atmosphere Project at the National Center for Atmospheric Research (NCAR) in Boulder, Colorado. At the same time, he served as a consultant to the Aeronomy Laboratory of the Environmental Research Laboratories of the National Oceanic and Atmospheric Administration. In 1977, Crutzen was appointed Senior Scientist and Director of the Air Quality Division at NCAR. From 1976 to 1981, Crutzen was also an adjunct professor at the Atmospheric Sciences Department of Colorado State University. In 1980, Crutzen returned to Europe, where he became affiliated with the Max-Planck-Institute for Chemistry in Mainz, Germany. From 1983 to 1985, he served as Executive Director of the Institute. Since 1987, Crutzen has also served as a part-time professor in the Department of Geophysical Sciences at the University of Chicago and at the Scripps Institute of Oceanography of the University of California at San Diego.

Robert F. Curl, Jr.

Curl received the Nobel Prize for Chemistry in 1996, along with Richard Smalley and Harold Kroto, for his discovery of the form of carbon known as buckminsterfullerene, or as buckyballs or fullerenes. This allotrope of carbon consists of soccer- or rugby-ball shaped molecules of 60 or more carbon atoms. The discovery of the fullerenes set into motion a dramatically new direction for chemical research that has led to amazing nanostructures with apparently unlimited applications in material sciences, electronics, and other fields.

Curl earned his B.A. in chemistry at Rice University in 1954 and a Ph.D. in chemistry at the University of California at Berkeley in 1957. After a year at Harvard University, Curl returned to Rice, where he was appointed Assistant Professor of Chemistry. Except for sabbatical leaves, he has remained at Rice ever since, becoming a full professor in 1967 and serving as Chair of the Department of Chemistry from 1992 to 1996.

Bassil I. Dahiyat

In 1997, Dahiyat coauthored an important paper with Stephen L. Mayo on the automated design of a nonnatural protein structure that was able to fold in essentially the same way that natural proteins fold. The computer program developed by Dahiyat and Mayo sorted through more than 10^{27} possible amino acid combinations to find a primary sequence that would cause the protein to fold in a manner similar to the folding found in a naturally occurring protein known as a zinc finger.

Dahiyat earned his B.S. and M.S.E. in biomedical engineering at Johns Hopkins University in 1990 and 1992, respectively. He then received his Ph.D. in chemistry at the California Institute of Technology in 1997.

Paul Dowd

Dowd published an important paper in 1995 describing the way in which vitamin K functions to promote the clotting of blood. That paper elucidated the process by which carbon dioxide released during the vitamin K cycle is used to carboxylate a variety of chemical species during the blood-clotting cascade.

Dowd received his A.B. from Harvard College in 1958, and his M.A. and Ph.D. from Columbia University in 1959 and 1962, respectively. Dowd was a research fellow (1962–63), instructor (1963–66), lecturer (1966–67), and an assistant professor (1967–1970) at Harvard University before joining the faculty at the University of Pittsburgh. At Pittsburgh, he served as Associate Professor of Chemistry until he was promoted to Professor of Chemistry in 1975.

At the time of his death in 1996, Dowd was pursuing four major lines of research: the mechanism of vitamin K action, the mechanism of vitamin B_{12} action, the expansion of rings in organic synthetic processes, and the study of trimethylenemethane. He was author or coauthor of more than 130 papers dealing with these and a variety of other topics.

Laura L. Dugan

Dugan was lead author of a 1997 paper on the use of carboxyfullerenes as agents for the protection of nerve cells against damage. She was awarded her S.B. in life sciences from the Massachusetts Institute of Technology and her M.D. from the Ohio State University School of Medicine. She did her medical residency training at Children's Hospital in San Francisco, an affiliate of the University of California at San Francisco. After leaving San Francisco, she worked as a postdoctoral fellow in the laboratory of Dennis Choi at Stanford University.

In 1993, Dugan joined the faculty at Washington University in St. Louis as an instructor in Neurology and Medicine, where she was promoted to Assistant Professor in 1995. She was awarded a Dana Research Fellowship and a Hartford Foundation Award for Geriatric Research at Stanford, and she received the Kopolow Award for Geriatric Psychiatry and Neurology at Washington University.

One area of special interest to Dugan is the role of free radicals in damage that occurs to the nervous system and in chronic nervous disorders, such as Alzheimer's disease and amyotrophic lateral sclerosis (ALS, or Lou Gehrig's Disease).

R. Michael Garavito

In 1995, a research team led by Garavito reported that they had determined the mechanism by which a derivative of aspirin binds to and inactivates an enzyme known as PGHS-1. This discovery holds promise for improving and expanding the range of drugs currently available for the treatment of pain. This research was part of an ongoing research effort aimed at trying to determine how the ability of aspirin to reduce pain and swelling can be explained on the basis of molecular structures.

Garavito has served as Associate Professor in the Department of Biochemistry at Michigan State University since 1995. He earned his B.A. in biology and anthropology in 1974 from the University of California at San Diego and his Ph.D. in biophysics and biochemistry in 1978 from Purdue University.

After completing his undergraduate studies at the University of California at San Diego, Garavito spent four years as a Graduate Research Assistant at Purdue while pursuing his doctoral studies. He then traveled to Basel, Switzerland, where he spent two years as a Postdoctoral Research Associate at the Biocenter. He remained at Basel for six more years, first as an independent research scientist at the Biocenter, and then as a Privat Dozent (lecturer) at the University of Basel.

In 1987, Garavito accepted an appointment as Assistant Professor in the Department of Biochemistry and Molecular Biology at the University of Chicago, where he was promoted to Associate Professor in 1994. In 1995, he moved to Michigan State.

Peter M. Gehring

In 1997, a research team on which Gehring participated reported on some interesting structural properties and phase transitions of the cubane molecule.

Gehring received his A.B. in physics and mathematics from Cornell University in 1982 and his Ph.D. in physics from the University of Illinois at Urbana-Champaign in 1989. After completing his doctoral program, he worked for three years as a research assistant and then assistant physicist in the Neutron Scattering Group at the Brookhaven National Laboratory. He also spent five months as a visiting scientist at the Laboratorie Leon Brillouin at Saclay, France.

In 1992, Gehring accepted a temporary appointment in the Reactor Radiation Division at the National Institute of Standards and Technology. Two years later, he was promoted to a permanent position as staff physicist, a post he currently holds.

M. Reza Ghadiri

In 1996, Ghadiri and his colleagues reported on a proteinlike polypeptide capable of self-reproduction. A year later, Ghadiri also reported on the development of an efficient new kind of biosensing device.

Ghadiri earned his B.A. from the University of Wisconsin at Milwaukee in 1982 and his Ph.D. in chemistry from the University of Wisconsin at Madison in 1987. He then spent two years in a postdoctoral program (1987–1989) at the Rockefeller University in New York City. He was an assistant professor at the Scripps Research Institute in La Jolla, California (1989–94), before being appointed Associate Professor in the Departments of Chemistry and Molecular Biology at Scripps. He also serves as a researcher at the Skaggs Institute for Chemical Biology at Scripps.

He has published more than three dozen papers on these topics and has been granted three patents on metallopeptides. Ghadiri has also been an invited lecturer at conferences, meetings, and other presentations in Japan, Mexico, Canada, Iceland, Germany, Spain, Switzerland, and Italy, as well as throughout the United States. His most recent honor was the 1995 American Chemical Society Award in Pure Chemistry. His research interests include the *de novo* design of functional artificial proteins, peptide nanotubes, and related biomaterials; artificial transmembrane ion and molecular channels; the design of biosensor arrays; construction of self-replicating molecular structures; and the study of early events in protein folding.

Russell J. Hemley

In 1997, Hemley reported on the superconductivity of sulfur at temperatures between 10 K and 17 K. This observation was important because it was the highest temperature at which any element had been shown to be superconductive.

Hemley received his B.A. in chemistry from Wesleyan University in 1977 and his M.A. and Ph.D. in chemistry from Harvard University in 1980 and 1983, respectively. He then stayed on at Harvard for another year as a postdoctoral fellow.

From 1984 to 1987, Hemley was a Carnegie Fellow at The Geophysical Laboratory (TGL) of the Carnegie Institution of Washington. He was then appointed Staff Scientist at TGL, a post he continues to hold. In addition to his work at TGL, Hemley has also served as Visiting Professor at the Johns Hopkins University (1991) and at the Ecole Normale Supérieure in Lyon, France (1996).

Steve Howdle

In 1996 and 1997, Howdle was a member of a research team that reported on breakthroughs in the use of supercritical carbon dioxide as a solvent for polar and ionic substances as well as a reaction medium for such materials.

Howdle earned his B.Sc. with honors in chemistry from Victoria University of Manchester, U.K., in 1986. He then continued his work at the University of Nottingham, from which he received his Ph.D. in 1989. He remained at Nottingham as a postdoctoral Venture Research Fellow from 1989 to 1991. He was then appointed Royal Society University Research Fellow in the Department of Chemistry at Nottingham, a post he continues to hold. Howdle's area of special interest is the application of supercritical fluids to chemical reactions and materials processing.

Sumion Iijima

In 1991, Iijima first observed the presence of carbon nanotubes in the soot produced during the electric arc vaporization of carbon. At the time, he was employed as a research fellow at the NEC Corporation in Japan.

Iijima received his undergraduate education at the University of Electro-Communications in Tokyo. He then transferred to Tohoku University, in Sendai, Japan, where he earned a Ph.D. in physics. He worked for 12 years (1970–1982) in the field of high-resolution electron microscopy at Arizona State University, first as a postdoctoral student, and later as a research scientist. Iijima also spent a year at Cambridge University in 1979 working on the structure of graphite as revealed through electron microscopy.

In 1982, Iijima returned to Japan to join the ERATO Ultrafine Particles Project. Five years later, he was appointed Research Fellow at NEC. Among Iijima's many honors are the 1996 Asahi Award, the Hishina Memorial Award, and the B. E. Warrne Award.

Chaitan Khosla

In 1997 a research team headed by Khosla found a method to take over the natural process by which an important antibiotic, erythromycin, is manufactured. That method can then be used to produce a whole new group of erythromycin-like antibiotics.

Khosla earned his bachelor's degree in chemical engineering at the Indian Institute of Technology in Bombay in 1985 and his Ph.D. in chemical engineering from the California Institute of Technology in 1990. After receiving his doctorate, Khosla was appointed Ramsay Memorial Fellow for Chemical Research in the Department of Genetics at the John Innes Institute in Norwich, U.K. (1990–92). He was then appointed Assistant Professor in the Department of Chemical Engineering at Stanford University, a post which he continues to hold. By courtesy, he was also appointed Assistant Professor in the Departments of Chemistry and Biochemistry at Stanford.

Eric T. Kool

In 1997, Kool and his colleagues reported on studies that indicate that the shape of nitrogen bases in the replication of a DNA molecule are at least as important as the hydrogen bonds formed between bases in such a molecule. If confirmed, these results would represent a dramatic breakthrough in our understanding of the mechanism by which DNA molecules make copies of themselves.

Kool's current positions are Professor of Chemistry at the University of Rochester, Associate Professor in the Department of Biochemistry and Biophysics, and Professor of Oncology and Member of the Cancer Center at the University of Rochester School of Medicine. He previously served as Associate Professor (1995–97) and Assistant Professor of Chemistry at Rochester (1990–95).

Kool received his B.S. in chemistry, *magna cum laude* and with honors in chemistry, from Miami University in 1982. He then earned his M.A., M.Phil., and Ph.D. from Columbia University in 1988, and he did his postdoctoral studies at the California Institute of Technology from 1988 to 1990. In the past decade, Kool has published more than 50 papers and given more than 70 invited talks at universities and industries located across the United States.

Harold Kroto

In 1985, Kroto, Richard E. Smalley, and Robert F. Curl, Jr., combined in the discovery of a new allotropic form of carbon known as buckminsterfullerene, a soccer-ball-shaped sphere consisting of 60 carbon atoms joined to each other in pentagons and hexagons. For this work, Kroto, Smalley, and Curl were awarded the 1996 Nobel Prize for Chemistry. In the same year, Kroto was knighted by Queen Elizabeth II.

Kroto received his bachelor's degree and his Ph.D. in chemistry from the University of Sheffield, U.K, in 1961 and 1964, respectively. After two years of postdoctoral research at the National Research Council in Ottawa, Canada, and one year at Bell Laboratories in New Jersey, Kroto accepted an appointment at the University of Sussex in Brighton in 1967. He became a full professor at Sussex in 1985 and, in 1991, was appointed Royal Society Research Professor at the University.

Paul Loubeyre

In 1995, Loubeyre reported on the formation of an unusual form of matter produced when oxygen and hydrogen react with each other at high pressure. The two gases did not combine explosively with the formation of water, as is generally the case, but, instead, formed an alloylike mixture of the two gases in the solid state.

Loubeyre attended the Ecole Normale Superieure in Paris from 1978 to 1982, where he wrote his university thesis. He then received his Doctorat es sciences Physiques in 1987 from the University of Paris and worked in a postdoctoral program for one year at the University of Illinois at Champaign-Urbana. Loubeyre then returned to Centre National de la Recherche Scientifique (CNRS) and the University of Paris, where he has been a research fellow at the two institutions ever since. Loubeyre was awarded the Bronze Medal of the CNRS in 1988, and the ESRF Young Scientist Award in 1996.

Jean-Pierre Majoral

In 1997, Majoral was leader of a research team that reported on the development of a new form of dendrimer in which reactive groups are contained within the structure of the molecule.

Majoral received the degrees of Maîtrise de Chimie (1964), Doctorat de 3ème cycle (1967), and Docteur ès-Sciences Physiques (1972) from Toulouse University, France. During the 1972–73 academic year, Majoral did his postdoctoral studies at the University of East Anglia in Norwich, U.K. He has held the posts of Assistant Professor at Toulouse University (1966–67), Attaché de Recherche at CNRS in Toulouse (1967–72), and Chargé de Recherche, Maître de Recherche, and Director of Research, second and then first class, at CNRS in Toulouse.

Majoral has published over 225 papers and 90 communications and has been invited to more than 60 professional conferences in France and more than 120 conferences in other nations throughout the world. In 1990, he was given the Award of the French Chemical Society by the Division of Coordination Chemistry.

Stephen L. Mayo

In 1997, Mayo and a graduate student, Bassil I. Dahiyat, reported on the development of a computer program that was able to predict the design of a simple protein using a few simple fundamental parameters about the general characteristics of amino acids and protein folding. The report has been regarded as a major step forward in the effort to better understand the rules by which proteins fold and how to apply those rules to the design and production of abiotic proteins.

Mayo's current positions are Assistant Professor of Biology in the Division of Biology at the California Institute of Technology and Assistant Investigator in the Structural Biology Section at the Howard Hughes Medical Institute. He also holds an appointment as Adjunct Assistant Professor of Biochemistry and Molecular Biology at the University of Southern California School of Medicine.

Mayo received his B.S. in chemistry, with distinction and honors, from Pennsylvania State University in 1983. In 1987, he received his Ph.D. in chemistry from the California Institute of Technology. In 1988, he accepted a one-year appointment as Miller Research Fellow at the University of California at Berkeley. He then spent a year working as Vice President for Biological Sciences at Molecular Simulations, Inc., a company he cofounded in 1984. He accepted his current appointment at Caltech in 1992, and he added his present appointment at the Howard Hughes Medical Institute in 1994.

Mario J. Molina

Molina was awarded a share of the 1995 Nobel Prize for Chemistry with F. Sherwood Rowland and Paul Crutzen. This award was based on the work done by these three men on the effects of chlorofluorocarbons (CFCs) on ozone in the Earth's upper atmosphere. In his research, Molina had discovered that CFCs, compounds that are normally very stable in the Earth's lower atmosphere, tend to break apart in the upper atmosphere under the influence of solar radiation. Some products of the decomposition of CFCs may attack ozone molecules, breaking them down to diatomic oxygen atoms. The environmental implications of these findings are highly significant since ozone in the stratosphere acts as a protective layer, shielding the Earth's surface from a large fraction of the Sun's deadly ultraviolet radiation.

Molina became interested in chemistry, he says, partly because of the influence of his aunt, who was a chemist. He was sent to a private school in Switzerland at the age of 11, and then returned to his native Mexico to complete his precollege education in 1959 at the Academia Hispano

Mexicana in Mexico City. He then entered the Universidad Nacional Autonoma de Mexico (UNAM), from which he earned his degree in chemical engineering in 1965.

Between 1965 and 1968, Molina spent two years at the University of Freiburg, Germany, studying on his own in Paris, and working at UNAM to develop the school's first graduate program in chemical engineering. Then, in 1968, he enrolled at the University of California at Berkeley to study physical chemistry.

At Berkeley, Molina studied under the noted research chemist and teacher, George Pimentel. He was awarded his Ph.D. in physical chemistry in 1972 and then remained at Berkeley for another year to continue his research in chemical dynamics with Pimentel. Finally, in the fall of 1973, Molina transferred to the University of California at Irvine, where he became a postdoctoral fellow in a research group headed by F. Sherwood Rowland. Rowland assigned Molina the task of working out the environmental fate of the CFCs, a project that lead to Molina's Nobel Prize–winning discoveries.

Molina left Irvine in 1982 to take a position as research scientist at the Jet Propulsion Laboratory in Pasadena, California. Six years later, he accepted a joint appointment as Professor in the Departments of Chemistry and of Earth, Atmospheric, and Planetary Sciences at the Massachusetts Institute of Technology (MIT). He currently holds the title of Martin Professor of Atmospheric Chemistry at MIT.

Jeffrey S. Moore

In 1997, a research team under Moore's direction reported on the construction of an abiotic oligomer with the ability to fold itself into a pattern similar to that found in naturally occurring zinc fingers. The research represents an important step forward in the understanding of the mechanisms that control protein folding as well as in the development of synthetic, proteinlike polypeptides.

Moore's current titles are Beckman Institute Professor and Professor of Chemistry and Materials Science and Engineering at the University of Illinois at Urbana-Champaign. His previous positions include Assistant Professor of Chemistry at the University of Michigan (1990–93) and Associate Professor of Chemistry and Materials Science and Engineering at Illinois (1993–95). Moore received his B.S. and Ph.D. from Illinois in 1984 and 1989, respectively.

Moore's special research interests include the study of molecular self-assembly and self-organization, energy and electron transfer in dendritic macromolecules, and structure-controlled macromolecules. He is the author of more than 80 publications and has presented invited lectures around the United States and in a number of foreign countries, including Belgium, Japan, the Netherlands, and France.

William J. Nellis

In 1997, Nellis and his colleagues reported the discovery of a metallic form of hydrogen produced by shock-compression methods developed in his laboratory. This discovery confirmed that hydrogen could exist in a form that had long been suspected, but never actually observed.

Nellis received his B.S. in physics from Loyola University, Chicago, in 1963, and his M.S. and Ph.D. in physics from Iowa State University in 1965 and 1968, respectively. After completing his doctoral studies, he accepted a position as Postdoctoral Research Associate in the Materials Science Division at the Argonne National Laboratory. He then served as Assistant Professor of Physics at Monmouth College (1970–73), before he accepted a position at the Lawrence Livermore National Laboratory, where he has been employed ever since. Between 1984 and 1994, Nellis was also Head of the Center for High Pressure Sciences of the University of California Institute of Geophysics and Planetary Physics. Nellis's research interests have focused on the study of condensed matter both during and after high-pressure shock compression.

Nellis has given invited talks at more than two dozen universities and national laboratories and has authored more than 150 articles in journals and conference proceedings.

Thomas V. O'Halloran

In 1997, O'Halloran and members of his research team reported on the role of certain "chaperone" proteins in the transport of copper across cell membranes. This discovery provided a better understanding of the way in which certain kinds of atoms and molecules move into and out of cells.

O'Halloran earned his B.S. in chemistry and his M.A. in bioinorganic chemistry from the University of Missouri at Columbia in 1979 and 1980, respectively. He then attended Columbia University in New York City, from which he received his Ph.D. in bioinorganic chemistry in 1985. O'Halloran did his postdoctoral work at the Massachusetts Institute of Technology from 1984 to 1986. His current position is Dow Professor at Northwestern University. Prior to this appointment, O'Halloran served as Professor (1994–), Associate Professor (1991–94), and Assistant Professor (1986–91) of Chemistry at Northwestern.

O'Halloran's special field of interest involves the biochemical roles of essential transition metals, such as zinc, iron, copper, and cytotoxic metals, such as platinum, cadmium, and mercury. He is also involved in research on DNA structure and the coordination chemistry and NMR (nuclear magnetic resonance) spectroscopy of mercury thiolate chemistry. In addition, he has reported on the earliest example of a metalloregulatory protein, a metal-ion receptor that controls gene expression, and on a novel mechanism for transcriptional regulation.

Hee-Won Park

In 1995, Park directed a team of researchers that reported on the crystal structure of the enzyme DNA photolyase from the bacterium *Escherichia coli*.

Park is currently at the St. Jude Children's Research Hospital in Memphis, Tennessee. He had previously worked as a student worker in the College of Veterinary Medicine at the Seoul National University in South Korea and as a research assistant and a teaching assistant in the Department of Biological Science at East Tennessee State University.

Park holds a D.V.M. from the School of Veterinary Medicine of Seoul National University (1985) and a Ph.D. in biochemistry from the University of Texas Southwestern Medical Center at Dallas (1995). He did his postdoctoral training, under the supervision of Lorena Beese, in the Department of Biochemistry at the Duke University Medical Center. Park currently has four publications to his credit, all dealing with the molecular structure of compounds.

George N. Phillips, Jr.

Phillips's special research interest involves the determination of the three-dimensional structure of protein molecules and the relationship between these structures and their biological functions. In 1996, he was part of a research team that reported on the structure of green fluorescent protein and the mechanisms by which that molecule emits light.

Phillips's current position is Professor of Biochemistry and Cell Biology at Rice University. He also holds posts as Adjunct Professor of Biochemistry at the Baylor College of Medicine and Scientific and Training Director at the W. M. Keck Center for Computational Biology. He received his B.A. in biochemistry and chemistry and his Ph.D. in biochemistry from Rice in 1974 and 1976, respectively. He did his postdoctoral work in structural biology at Brandeis University (1977–82).

Phillips's first teaching position was as Assistant Professor of Biophysics and Biochemistry at the University of Illinois (1982–87). He then joined the Rice faculty as Associate Professor of Biochemistry (1987–92) before being promoted to his current position.

Paul D. Ronney

Ronney served as primary ground controller for a series of experiments carried out during Spacelab mission MSL-1 on space shuttle flight STS-83 in April 1996. He had been selected by the National Aeronautics and Space Administration (NASA) to train as an Alternative Payload Specialist for Spacelab mission MSL-1, but was not chosen to fly on the shuttle itself. Ronney's special assignment was a group of experiments called SOFBALL (Structure of Flame Balls at Low Lewis-number) designed to study flame structure and properties in microgravity environments.

Ronney holds a B. S. in mechanical engineering from the University of California at Berkeley (1978), a M. S. in aeronautics from the California Institute of Technology (1979), and a D.S. in aeronautics and astronautics from the Massachusetts Institute of Technology (1983).

After earning his degree at MIT, Ronney held postdoctoral appointments at NASA's Lewis Research Center (1983–85) and at the U.S. Naval Research Laboratory (1985–88). He then served as Assistant Professor in the Department of Mechanical and Aerospace Engineering at Princeton University (1986–93). In 1993, he moved to the University of Southern California, where he holds a joint appointment as Associate Professor in the Departments of Mechanical Engineering and Aerospace Engineering.

Antoine Rousse

Rousse's primary area of research interest is in femtosecond X-ray diffraction studies of organic and protein materials. These studies identify changes in molecules that take place on a scale of quadrillionths of a second. In 1997, a team headed by Rousse reported on developments in techniques for making such measurements.

Although Rousse was originally interested in playing professional basketball and even played on the French National Championship team in 1983, he eventually turned to physics and earned his D.E.A. and Ph.D. in plasma physics from the University of Orsay in Paris in 1990 and 1994, respectively.

In the years before earning his Ph.D., Rousse also worked for the IBM Corporation on the etching of semiconductors and on the study of nanosecond experiments in hot plasmas. After completing his degree, Rousse spent a year at RIKEN (Institute of Physical and Chemical Research) in Tokyo working with subpicosecond high-intensity lasers, coherent X-ray and UV production, and solid material processing. Since 1995, he has been senior researcher at LOA in Palaiseau, France.

F. Sherwood Rowland

Rowland was awarded a share of the 1995 Nobel Prize for Chemistry with Mario Molina and Paul Crutzen. The prize was awarded for work done by these three scientists on the effects of chlorofluorocarbons (CFCs) on the ozone layer of the Earth's atmosphere.

Rowland was awarded his A.B. from Ohio Wesleyan University in 1948 and then transferred to the University of Chicago for graduate study. At Chicago, Rowland earned his M.S. in chemistry in 1951 and his Ph.D. in chemistry the following year.

In the fall of 1952, Rowland accepted an appointment as instructor in chemistry at Princeton University, a post he held for four years. He then accepted a position as Assistant Professor of Chemistry at the University of Kansas, where he was promoted to Associate Professor in 1958 and Professor in 1963.

In August of 1964, Rowland was appointed Professor of Chemistry and first Chair of the Department of Chemistry of the new University of California campus at Irvine, a campus that was not to open until a year later. At Irvine, Rowland continued his research on radiochemistry, or "hot-atom chemistry," that had been supported at Kansas by the Atomic Energy Commission (AEC). Support from the AEC and its successors (the Energy Research and Development Administration and, later, the Department of Energy), continued to make Rowland's research possible for the next three decades.

In 1970, Rowland retired as Chair of the Department of Chemistry and became interested in the environmental chemistry of the chlorofluorocarbons (CFCs). The CFCs had been discovered in the 1920s and had become widely popular for a variety of commercial and industrial applications. One of the reasons for their popularity was that they were very stable under normal atmospheric conditions near the Earth's surface.

What Rowland and his graduate student, Mario Molina, soon discovered, however, was that CFCs were far less stable under conditions present in the upper atmosphere. In these regions, energy from the Sun caused CFCs to break apart, producing species that attack and destroy ozone molecules. This problem has severe environmental consequences, since ozone in the stratosphere acts as a protective shield that prevents harmful ultraviolet radiation from the Sun from reaching the Earth's surface. It was for this discovery that Rowland, Molina, and Crutzen were awarded the 1995 Nobel Prize for Chemistry.

Matthias Schädel

One of Schädel's areas of special interest has been the development of techniques for separating and analyzing super-heavy elements produced during nuclear reactions in particle accelerators. This research area has posed particularly demanding challenges, since elements produced by this process typically have half-lives of only a few seconds or less. In early 1997, a research team under Schädel's direction reported a remarkable series of experiments in which the transfermium element seaborgium (number 106) was studied and its position in the periodic table was compared with that of other elements.

Schädel is the lead nuclear chemist at the Gesellschaft für Schwerionenforschung (GSI), the heavy-ion laboratory, at Darmstadt, Germany. The laboratory is one of the leading centers for research on transfermium elements and has been given credit for the discovery or a share in the discovery of elements 107 through 112.

Schädel received his diploma in chemistry in 1974 and a doctoral degree in nuclear chemistry from the University of Mainz in 1979. He joined the GSI research team in 1976 and has worked at the facility ever since. He has also participated in research at the U.S. National Laboratories Lawrence Livermore and Lawrence Berkeley.

Christine Schmidt

In 1997, Schmidt was lead author on an article dealing with the use of electrically conductive polymers to treat damaged nerve cells. This study was one of a series being carried out at various institutions to find ways of helping damaged nerves begin to regrow.

Schmidt earned her B.S. in chemical engineering at the University of Texas in 1988. She then received her Ph.D. in chemical engineering from the University of Illinois at Urbana-Champaign in 1994. Schmidt then spent two years as a postdoctoral fellow in chemical engineering at the Massachusetts Institute of Technology (MIT).

In 1988, Schmidt was appointed Summer Research Engineer at the Mobil Chemical Company in Edison, New Jersey, and between her undergraduate and graduate programs, she spent a year as a research scientist at the University of Texas in Austin. Upon completing her postdoctoral studies at MIT, she accepted an appointment as Assistant Professor of Chemical Engineering at the University of Texas.

Georg E. Schulz

In 1977, Schulz was lead author for a research group that reported on the three-dimensional structure of squalene cyclase, an enzyme responsible for a cyclization reaction in the formation of an important isoprenoid molecule. The study provided valuable information on the process by which squalene is synthesized in cells.

Schulz studied physics and engineering at the Technical University of Berlin (1958–62) and then physics at the Ruperto-Carola-Universität Heidelberg (1962–64). In 1966, he was awarded his Ph.D. in physics from Heidelberg.

After a year of postdoctoral study at Yale University (1967–68), Schulz returned to Germany, where he established a protein crystallography group at the Max-Planck-Institute at Heidelberg, and he continues to direct the work of that group. In 1973, Schulz received his Habilitation and Venia legendi in biophysics from the Ruperto-Carola-Universität at Heidelberg. Six years later, he was appointed Professor of Physics at Ruperto-Carola, a post he continues to hold. In 1984, Schulz was also appointed Professor of Biochemistry at the Albert-Ludwigs-Universität at Freiburg.

Mark D. Scott

In 1997, Scott was lead author of an article on the production of "stealth erythrocytes," red blood cells that are camouflaged in such a way as to allow them to enter human cells without being rejected by the body's immune system. The development of these cells provide a new mechanism for dealing with some of the problems experienced in certain kinds of blood transfusions.

Scott earned his B.A. and M.A. from Western State College in Gunnison, Colorado, in 1979 and 1984, respectively. After receiving his bachelor's degree, he served as a Rodent Control Biologist for the U.S. Peace Corps in Dakaro, Niger (1979–80). Later, he attended the University of Minnesota, from which he received his Ph.D. in 1988. For his postdoctoral work, Scott was a research fellow at the University of Minnesota's Department of Laboratory Medicine and Pathology (1984–87). He also held positions at the Children's Hospital Oakland Research Institute (1988–89) and the Institut National de la Santé et de la Recherche Médicale, Créteil, France (1989–90).

In 1993, Scott was appointed Assistant Professor in the Division of Experimental Pathology of the Department of Pathology and Laboratory Medicine at Albany Medical College in Albany, New York. He was later promoted to Associate Professor (1996) and to Director of Graduate Studies in the Division of Experimental Pathology (1996).

Tsung-Chung Shen

In 1995, a research group under the direction of Shen reported on new methods for breaking individual chemical bonds in a molecule. The methods involve the use of a scanning tunneling microscope (STM) to break bonds between silicon and hydrogen atoms in a semiconducting material.

Shen has served as Senior Research Scientist at the Beckman Institute of the University of Illinois at Urbana-Champaign since 1996. His previously held positions include Visiting Research Assistant Professor (1994–96) and Research Associate (1989–94) at the Beckman Institute; Affiliate Assistant Professor in the Department of System Science and Mathematics at Washington University (1988–89), and Adjunct Assistant Professor in the Department of Physics at the University of Missouri at St. Louis (1988–89).

Shen earned his B.S. in physics at the National Taiwan University in 1975 and his M.S. in physics at the National Tsing Hua University in 1977. He was awarded his Ph.D. by the University of Maryland in 1985.

Jens Christian Skou

Skou was awarded one half of the 1997 Nobel Prize in Chemistry, sharing that award with Paul D. Boyer and John E. Walker. Skou's award was in recognition of his discovery of an enzyme known as Na^+,K^+-ATPase, which is responsible for the conduct of sodium and potassium ions across nerve membranes. The imbalance between sodium and potassium ions had been known since the early 1920s. Three decades later, Richard Keynes and Alan Hodgkin demonstrated that the stimulation of a nerve cell causes the flow of sodium and potassium ions across a cell membrane. Then, in 1957, Skou showed precisely how such transport takes place as a result of the action of Na^+,K^+-ATPase.

Skou studied at the University of Copenhagen, Denmark, from which he received his *cand.med.* (M.D.) and *dr.med.* (Doctor of Medical Science) in 1944 and 1954, respectively. During the period between his receiving these two degrees, he was involved in clinical training at the Hjørring Hospital and Aarhus Orthopaedic Clinic and held posts as Assistant Professor and Associate Professor at the Institute of Physiology of the University of Aarhus. In 1963, he was appointed Professor and Chairman of the Institute. From 1978 to 1988, he served as Professor of Biophysics at the University of Aarhus.

Richard E. Smalley

Smalley shared the 1996 Nobel Prize in Chemistry with Robert Curl and Harold Kroto for the discovery in 1985 of a new allotropic form of carbon known as buckminsterfullerene, whose molecules are known as fullerenes or "buckyballs." Buckyballs consist of 60 carbon atoms joined by hexagons and pentagons in a closed, soccer-ball-like formation. The new allotrope was named in honor of the architect Buckminster Fuller, who developed the geodesic dome as a modern form of construction in the 1950s. Since 1985, Smalley has continued his research on the fullerenes and, most recently, pursued the development of nanoscale materials that can be made from fullerenes.

Smalley attended Hope College, Holland, Michigan, from 1961 to 1963 and then transferred to the University of Michigan, from which he received his B.S. in chemistry in 1965. He then earned his M.A. and Ph.D. from Princeton University in 1971 and 1973, respectively. In the period between 1965 and 1969, Smalley worked as a research chemist for the Shell Chemical Company.

Smalley's first academic appointment was at the University of Chicago, where he served from 1973 to 1976. He then moved to Rice University, where he has served as Assistant Professor (1976–80), Associate Professor (1980–81), Professor (1981–82), and the Gene and Norman Hackerman Professor of Chemistry (1982–). Since 1990, he has also held an appointment as Professor of Physics at Rice.

Laura Stern

In 1996, Stern was coauthor of a report on the superheating of ice trapped in methane clathrates. The report was of some interest because it provided one of the few instances in which water ice remained in the solid state at temperatures significantly above its normal melting point. In recognition of this discovery, Stern's research team was awarded the 1997 Richard A. Glenn Award from the Fuel Chemistry Division of the American Chemical Society.

Stern is a geophysicist with the U.S. Geological Survey (USGS) in Menlo Park, California. She received her B.S. and M.S. in geology from Stanford University in 1983 and 1985, respectively. She has been employed at the USGS since leaving Stanford in 1985. Her work at Menlo Park has focused on the properties of icy compounds and gas hydrates under both terrestrial conditions and conditions present on other planets. She has also studied the effects of hydrothermal reactions along fault lines and reactions occurring in sinking parts of the ocean crust.

Kenneth S. Suslick

Suslick is a leading authority on the effects of sound and ultrasound energy on chemical reactions. In 1997, a research team under his direction reported on the effects of high-pressure bubble implosion on the breaking of chemical bonds.

Suslick received his B.S. from the California Institute of Technology, with honors, in 1974 and his Ph.D. from Stanford University in 1978. From 1974 to 1975, he worked on laser isotope separations as a chemist at Lawrence Livermore Laboratory in Livermore, California.

In 1978, Suslick accepted an appointment as Assistant Professor at the University of Illinois at Urbana-Champaign. He was promoted to Associate Professor in 1984 and to Professor in 1988. He also currently holds a five-year appointment as the William H. and Janet Lycan Professor of Chemistry at Illinois. Previously he has held the posts of Professor at the Beckman Institute for Advanced Science and Technology at Illinois (1989–92), Alumni Research Scholar Professor of Chemistry at Illinois (1995–97), Professor of Materials Science and Engineering at Illinois (1993–), and Visiting Fellow at Balliol College and Inorganic Chemistry Laboratory at Oxford University (1986).

Reshef Tenne

In 1997, Tenne coauthored a paper describing the use of hollow fullerene-like particles made of tungsten disulfide that appear to have significant promise as the components of solid lubricants. His current position is Professor of Materials and Interfaces at the Weizmann Institute in Rehovot, Israel.

Tenne received his B.Sc. in chemistry and physics in 1969, his M.Sc. in physical chemistry in 1971, and his Ph.D. in theoretical chemistry in 1976, all from the Hebrew University in Jerusalem. His postdoctoral studies were carried out in theoretical chemistry at the Battelle Institute in Geneva, Switzerland, in 1978.

Tenne's previous appointments include Researcher in the Electrochemistry Group at Battelle (1979), Scientist in Photoelectrochemistry (1979), Senior Scientist in Photoelectrochemistry and Electronic Materials, and Associate Professor (1985) and Professor (1995) at the Weizmann Institute of Science. He has also served as Visiting Professor at the Solid State Institute, Technion, Haifa (1986); at the Pierre et Marie Curie University in Paris (1988); at the University of Tokyo (1991); and at the Solid State Physics Laboratory at CNRS (1989). In addition, from 1983 to 1986, he held the Helen and Milton A. Kimmerlman Career Development Chair at the Weizmann Institute.

Donald A. Tomalia

Tomalia is credited as being one of the pioneers in dendrimer research. In 1985, he reported on the preparation of a type of dendrimer since given the name of Starburst® Dendrimers, a type of molecule in which branches of the dendrimer that disperse outward form a central core whose appearance suggests the name that has been given them.

At the time, Tomalia was working as a Research Scientist at Dow Chemical Company. He had earlier served as Chemist (1962–64), Research Chemist (1967–67), Project Leader (1967–68), Group Leader (1968–71), Research Manager (1971–76), Associate Scientist (1976–79), and Senior Associate Scientist (1979–84) at Dow. A few years after his discovery of Starburst® Dendrimers, Tomalia left Dow to start up his own company. His company, Dendritech, is now a division of Michigan Molecular Institute (MMI), a chemistry "think tank" located in Midland, Michigan. Tomalia was also cofounder of Oxazogen, Inc., and is the Director of Nanoscopic Chemistry and Architecture and a research professor at MMI.

Tomalia earned his B.A. from the University of Michigan at Flint in 1961, his M.S. from Bucknell University in 1962, and his Ph.D. from Michigan State University in 1968. In addition to his corporate positions, he has been an adjunct professor at Central Michigan University and at Michigan Technological University since 1993.

Yoshiharu Waku

In 1997, Waku and his coworkers reported on the development of a ceramic material that remains flexible at high temperatures. This development has very significant practical potential since flexibility is one of the few properties that ceramics do not normally possess and which, if present, would greatly expand the variety of uses for ceramics.

Waku received his undergraduate degree from the Mineral Department of Akita University, Japan, in 1970. He then received his master's and doctoral degrees from the Graduate School of Engineering at Tohoku University in 1975 and 1979, respectively. Since 1980, Waku has been employed at Ube Industries, where his research involves the study of composite materials, fracture mechanics, and amorphous materials.

John E. Walker

Walker shared the 1997 Nobel Prize in Chemistry with Paul D. Boyer and Jens C. Skou. One half of the prize was shared by Walker and Boyer for their explanation of the process by which adenosine triphosphate (ATP) is synthesized in cells. ATP is one of the most important molecules in living cells because of its ability to transport the energy needed to drive many essential biochemical reactions. ATP is manufactured in cells from a simpler compound, adenosine monophosphate (AMP), by the stepwise addition of two phosphate groups. This process is catalyzed by an enzyme known as ATP synthase.

Walker's contribution to this research was to discover the precise structure of ATP synthase and to demonstrate how this structure could support Boyer's theory. Walker first determined the primary structure of the protein (its amino acid sequence) and then its secondary and tertiary structures. With this information, he was able to construct a molecular model for the enzyme that validated Boyer's hypothesis about the function of ATP synthase.

Walker attended St. Catherine's College at the University of Oxford, from which he received his B.A. in chemistry in 1964. Walker then spent four years as a research student at the Sir William Dunn School of Pathology at Oxford before earning his M.A. and D. Phil. from Oxford in 1969.

Walker then spent two years as a postdoctoral fellow at the School of Pharmacy of the University of Wisconsin at Madison. From 1971 to 1974, he was appointed NATO Fellow at CNRS in Gif-sur-Yvette, France, and EMBO Fellow at the Institut Pasteur in Paris. In 1974, Walker accepted an appointment as a member of the scientific staff at the Cambridge Laboratory of Molecular Biology of the Medical Research Council. He was assigned to the Division of Protein and Nucleic Acid Chemistry at the laboratory. In 1982, Walker was promoted to Senior Scientist at the laboratory and, in 1987, he was given a special appointment (professorial grade) at the laboratory.

Robert M. Wightman

Wightman coauthored a 1995 paper that described the development of a system for identifying individual molecular reactions in solution. The technique that he developed, in conjunction with Maryanne M. Collinson, holds promise for providing chemists with a way of studying the details of chemical reactions that had never been available before.

Wightman received his B.A. from Erskine College in South Carolina in 1968. He received his Ph.D. in analytical chemistry from the University of North Carolina at Chapel Hill in 1974. Wightman then spent two years as a postdoctoral student at the University of Kansas (1974–76).

Upon completion of his academic training, Wightman accepted an appointment as Assistant Professor of Chemistry at Indiana University, where he served until 1989. He was promoted to Associate Professor in 1982 and to Professor in 1985. In 1989, Wightman was appointed the S.

R. Kenan, Jr., Professor of Chemistry at the University of North Carolina at Chapel Hill, a post he currently holds. Also, in 1984, Wightman accepted a six-month appointment as Research Associate at the London Hospital Medical College of the University of London in his native England.

Taner Yildirim

During the late 1990s, Yildirim was a member of a research team working on the structure and properties of cubane. Cubane is of unusual interest to chemists and physicists because it is the most highly strained, kinetically stable ring system that can be prepared in quantities large enough to study. The compound promises to have a number of important applications as explosives, specialty polymers, fuel additives, and antiviral agents.

Yildirim attended the Middle East Technical University in Ankara, Turkey, earning his B.S. and M.Sc. in physics in 1988 and 1990, respectively. In 1990, Yildirim came to the United States for his doctoral studies and was awarded his Ph.D. in physics from the University of Pennsylvania in 1994. He has been a research associate at the University of Maryland in College Park and at the National Institute of Standards and Technology (NIST) Center for Neutron Research in Gaithersburg, Maryland, since 1994.

Masasuke Yoshida

Yoshida is the senior member of a research team from the Tokyo Institute of Technology and the Faculty of Science and Technology of Keio University that reported in 1997 on the rotation of the gamma subunit within the ATP synthase molecule. The report was particularly important and interesting because of its association with research on the enzyme that won the Nobel Prize for Paul Boyer and John Walker later in the same year.

Yoshida received his bachelor's and doctoral degrees from Tokyo University in 1966 and 1972, respectively. He then worked as a research associate at Jichi Medical School in Tokyo (1972–79) and at the University of California in San Diego (1979–81). He returned to Jichi for two years to serve as Associate Professor of Chemistry before accepting an appointment as Associate Professor of Chemistry at the Tokyo Institute of Technology (TITech). He was promoted to Professor of Chemistry at TITech in 1989.

J. S. Zambounis

Zambounis is part of a research team at Ciba Specialty Chemicals that reported in 1997 on the development of a soluble pigment. The discovery was important because it provides a new mechanism by which otherwise insoluble colorants can be applied to surfaces.

Zambounis earned a bachelor's degree in chemistry at the University of Athens, Greece, in 1984 and then continued his studies at the National Hellenic Research Foundation. He received his Ph.D. from the Foundation in 1987 for his study on organic conductors.

In 1988, Zambounis accepted a position at the Central Research Laboratories of Ciba-Geigy AG in Basel, Switzerland. At Ciba-Geigy, he continued his research on organic conductors and superconductors. In 1992, he moved to Corporate Materials Research at Ciba-Geigy, where he was placed in charge of a series of projects on pigments. In 1995, Zambounis accepted a new position with Marketing Center Paints of the High Performance Pigments Division of Ciba Specialty Chemicals. In this post, he is in charge of marketing and applications of organic pigments for decorative paints and wood finishes.

CHAPTER FIVE
Documents and Reports

C hapter 3 of this book discussed the close interaction between chemical research and chemical engineering and other facets of human life. Chemists are constantly devising new materials that change the way we live our lives, the problems we face, and even our philosophical outlook on the world. It should hardly be surprising that strong differences of opinion would arise as to how we are to deal with advances in chemistry and chemical engineering.

This chapter contains statements that have been made about controversial issues involving chemical research. The statements have been arranged according to topic in an outline similar to that used in chapter 3.

DETERIORATION OF THE OZONE LAYER

Some of the most important science-based problems facing the world today involve an element of speculation. The issues of global climate change and ozone depletion are examples. Measuring changes in the concentration of carbon dioxide or ozone in the atmosphere and charting those changes over many years or decades is a difficult task. Scientists are often hard-pressed to say whether they can confirm long-term trends in warming, loss of ozone, or other important changes. It is perhaps even more difficult to say to what extent human activities are responsible for changes that are observed or what the effect of those changes may be over the next decade or the next century.

Such issues, however, can have profound effects on the ways humans conduct their everyday activities and the extent to which governments try to regulate those activities in order to protect the Earth's natural resources. Under these circumstances, it should hardly be surprising that debates over scientific data, the significance of those data, and the actions that should be based on them can be vigorous and ongoing.

In 1995, the Subcommittee on Energy and Environment of the Committee on Science of the U.S. House of Representatives held hearings at which representatives of two very different views of the "ozone problem" presented their positions. Two members of the scientific "establishment" tried to present the case that ozone depletion was actually occurring and that governments needed to respond to the problem. A third scientist argued that this position was based on inaccurate scientific research and inappropriate extrapolations to political action. The statements below are taken from testimony offered at those hearings.

Reference: "Scientific Integrity and Public Trust: The Science behind Federal Policies and Mandates: Case Study 1—Stratospheric Ozone: Myths and Realities," Hearing before the Subcommittee on Energy and Environment of the Committee on Science of the U.S. House of Representatives, 104th Congress, 1st session, 20 September 1995.

Statement by Robert T. Watson, Ph.D.
Associate Director of Environment, Office of Science and Technology Policy, Executive Office of the President

It's a pleasure to be able to address what I believe to be a real success story—credible science combined with technological advances that have led to informed policy formulation at the national and international level.

The scientific community, industry, environmental organizations, and governments have all worked towards a common goal—the cost-effective protection of human health and our vital ecological systems.

The American public can be proud that the U.S. provided scientific and policy leadership, and partisan politics were put aside to protect the health of Americans.

My testimony represents the view of the very, very large majority of the international scientific community from academia, industry, government labs, and environmental organizations, not the view of single individuals with few, if any, relevant publications in the peer-reviewed journals.

Hundreds of scientists from developed and developing countries, some of whom at one time were skeptics, have been involved in the preparation and peer-review of these assessments.

I believe it's particularly important to note that industry scientists and industry-sponsored research played a vital role in these assessments.

The key issues are very simple. The ozone layer limits the amount of UV-B radiation reaching the Earth's surface. Thus, a decrease in ozone will lead to an increase in UV-B radiation reaching the Earth's surface. Increased levels of UV-B reaching the Earth's surface will, not may, have adverse consequences for human health, ecological systems, and air quality.

There is absolutely no doubt that the major sources of atmospheric chlorine are from human activities, not from natural sources. Human activity is also a major source of atmospheric bromine.

Photochemically active halogen species can catalytically destroy stratospheric ozone. Each chlorine molecule can destroy tens of thousands of ozone molecules, and bromine is at least 50 times more efficient.

Since the late 1970s, ground-based balloon and satellite data have documented significant decreases in column content of ozone over Antarctica, about 60 percent, as shown in one of my figures in my testimony, and drastic changes in the vertical distribution, close to 100 percent loss of ozone at certain altitudes.

The Antarctic ozone holes in 1990, 1992, 1993, and 1994, were the most severe on record.

As we speak today, and as expected, satellite, balloon, and ground-based data show that the Antarctic ozone hole is once again developing in the fashion similar to the last few years.

There is absolutely no doubt that the springtime Antarctic ozone hole is due to the increasing concentrations of anthropogenic chlorine and bromine. This conclusion is based on combining extensive ground, aircraft, balloon, and satellite data with laboratory data and theoretical modeling.

The speculative and totally unsubstantiated hypothesis of Dr. Singer presented before Congress a few weeks ago is totally inconsistent with the observational data and theory.

With respect to global ozone, the observational data, as I've shown in figure 4 [figures not reproduced here] of my testimony, provide conclusive evidence that ozone depletion is occurring at all latitudes, except the tropics, and in all seasons.

Analysis of extensive ground-based Dobson and TOMS data through 1994 has shown that column ozone has decreased by 5 to 6 percent in summer in the northern hemisphere, 9 to 11 percent in winter/spring in the northern hemisphere, 8 to 9 percent in southern mid-latitudes on a year-round basis.

Figure 5 in my testimony also shows the seasonal and latitudinal trends, illustrating the very significant trends at middle and high latitudes.

In each case, the natural periodic and episodic fluctuations are taken into account—solar cycle, season, and volcanic activities.

The weight of scientific evidence strongly suggests that the observed mid-latitude ozone trends are due in large part to anthropogenic chlorine and bromine.

Ozone depletion is expected to peak within the next year or so, reaching about 6 to 7 percent ozone depletion in northern mid-latitude in summer and fall over the USA, and 12 to 13 percent in winter over northern mid-latitudes, and about 11 percent in southern mid-latitudes.

The projected changes in column ozone would be accompanied by 15 percent, 8 percent, and 13 percent increases in surface erythermal radiation in winter/spring in the northern mid-latitudes, summer/fall at the northern mid-latitudes, and in the southern hemisphere year-round.

The link between a decrease in stratospheric ozone and an increase in surface UV has been further strengthened in recent years. Measurements in Antarctica, Australia, Canada, and Europe have shown under clear-sky conditions that when column ozone decreases, the amount of UV-B increases, exactly as expected by theory. . . .

Of particular importance for human health are the increases in the incidence of non-melanoma skin cancer, melanoma skin cancer, eye cataracts, and a possible suppression of the immune-response system.

Some, such as Fred Singer and Sallie Baliunas, try irresponsibly to trivialize the issue of ozone depletion by noting that an ozone depletion of the magnitude observed is equivalent to only moving south by 100 miles or so.

The reason this risk is even this low is the success of the Montreal Protocol and its Amendments and adjustments.

Without these international agreements, we would be facing future increases in UV-B radiation of possibly 40 to 50 percent by the middle or the end of the next century, and the comparable distance to move would be more like 1,000 miles or so.

There's a large difference in skin cancer rates between cities in the northern half of the U.S. and those in the southern half. The difference for white, Anglo-Saxon males in Albuquerque and Seattle is at least a factor of five difference.

In conclusion, the Montreal Protocol and its amendments and adjustments are a success story that will in the future save thousands of American lives each year. Who amongst us would want to turn back the clock by weakening the Montreal program, leading to the deaths of innocent Americans for the sake of a few dollars?

Statement by S. Fred Singer, Ph.D.
President, The Science and Environmental Policy Project

Examples of Failures of Scientific Integrity:

Today's hearing on scientific integrity as related to the stratospheric ozone layer is well-timed. The United Nations Environment Programme and the secretariat for the Montreal Protocol [on Substances that Deplete the Ozone Layer] designated September 16 as the first annual International Day for the Preservation of the Ozone Layer. The White House, spurred on by the EPA, has extended the celebration into a whole week. This should remind us that ozone depletion is no longer just a scientific debate; entrenched domestic and international bureaucracies, not to mention commercial interests, now have a considerable stake in keeping alive fears of an ozone catastrophe.

This morning, I will touch on several topics that relate to the theme of scientific integrity:

First, I want to state clearly that there is no scientific consensus on ozone depletion or its consequences. "Consensus" is a political concept, not a scientific one. It is used mainly to gain reassurance for an ideological position and to avoid having to examine the scientific arguments in detail. Consensus has been claimed also for the global warming issue. The official report from the UN-sponsored Intergovernmental Panel on Climate Change mentions the existence of "minority" views, but the editors could not, or perhaps would not "accommodate" them. The IPCC editors thus achieved "consensus" by ignoring contrary evidence and dissenting views. Much the same has been true in the ozone issue.

In view of the present policy to ban CFCs by the end of 1995, why spend a lot of energy fighting a *fait accompli*? I think the best answer was given by an environmental activist on an ABC News-"Nightline" television program in 1994. Michael Oppenheimer of the Environmental Defense Fund complained that "if [skeptical scientists] can get the public to believe that ozone wasn't worth acting on, that they [the public] were led in the wrong direction. . ., then there is no reason for the public to believe anything about any environmental issue." Given the activist groups' miserable record of unfounded scares about the global environment, such a reaction may be warranted.

Next, I want the record to show that the 1987 Montreal Protocol [on Substances that Deplete the Ozone Layer] was negotiated without adequate concern for scientific evidence. The chief U.S. negotiator, State Department official Richard Benedick, proudly revealed in his 1991 book, *Ozone Diplomacy*, on page 2: "Perhaps the most extraordinary aspect of the treaty was its imposition of substantial short-term economic costs . . . against unproved future dangers—dangers that rested on scientific theories rather than on firm data." Again, on page 18: "In July 1987, practically on the eve of the final negotiating session in Montreal, NOAA concluded that the 'scientific community is currently divided as to whether the existing data on ozone trends provides sufficient evidence . . . that a chlorine-induced ozone destruction is occurring.'"

Benedick does not mention the fact that, as late as 1988, published evidence on stratospheric chlorine showed no upward trend, thus indicating that neither CFCs nor other manmade chemicals were contributing significantly to the total—over and above known natural sources like volcanoes and oceans. An article by MIT professor Ronald Prinn, in a book edited by Professor Sherwood Rowland and published in 1988, makes this point quite clear.

It is apparent from the above quotes that the negotiators and their scientific supporters were not at all inhibited by the absence of scientific evidence—nor indeed by the presence of contrary information.

Third, the self-constituted Ozone Trends Panel first announced the existence of global ozone depletion in a March 1988 press conference, but did not present its supporting analysis for review until much later. A study of the OTP data by two independent American scientists, which was widely distributed as a preprint, showed clearly that, *even after they thought they had successfully "subtracted" the natural variations* by statistical methods, the so-called "depletion trend" depended on the choice of time interval—i.e., the year the analysis starts and ends. Curiously, this result, which shows the dominance of the large (natural) solar-cycle variation of ozone, was left out of a later published paper involving the same authors as collaborators with government scientists.

There is also a still unresolved dispute about the quality of the data themselves. The OTP, and the subsequent UN-sponsored assessment groups, have never grappled with objections published by two Belgian researchers in 1992. These scientists showed that the ozone readings were contaminated by air pollution and termed the reported ozone trend "fictitious." Because of similar absorption of ultraviolet, decreases in sulfur dioxide, brought about by reduced industrial emissions, were being falsely read as decreases in ozone.

Global ozone depletion is still a controversial subject. Starting with the OTP press conference, depletion has generally been reported to be "worse than expected." This statement should produce the logical conclusion that the CFC-ozone theory (on which "expectation" must be based) is wrong, or the observations are wrong, or they are both wrong.

Fourth, another press conference, arranged by NASA on February 3, 1992—during crucial Congressional hearings on the NASA budget and well before the end of the series of stratospheric observations—implied the threat of an Arctic ozone hole. The resulting nationwide scare led the Bush White House to advance the phase-out of CFCs to December 31, 1995.

The Arctic ozone hole never happened—something NASA scientists could have predicted at the time of the press conference. Information leaked to a journalist indicated that NASA scientists had mid-January satellite data showing that stratospheric chlorine was already in decline. Yet the agency went ahead with the February 3 press conference and refused to reveal this information and allay public fears until a second NASA press conference three months later, on April 30.

Fifth, the "smoking gun" of ozone depletion activists is, of course, an increasing trend of solar ultraviolet radiation at the Earth's surface. All of the published evidence before November 1993 had shown no such trend. Then, a research paper in *Science* magazine claimed upward trends of as much as 35 percent per year—without giving any estimate of the margin of error. This widely touted result, featured in a press release by *Science* and still being cited by the EPA and environmental activist groups, was shown to be completely spurious. The analysis was based on faulty statistics; the "trend" was zero.

Only later was it learned that the paper had been first submitted to the British scientific journal *Nature*, but had been rejected in the peer-review process. It's still somewhat of a mystery how this article passed the review process of *Science*.

There is still no evidence for an increased trend of surface UV to match a putative ozone depletion trend.

Finally, there is the Setlow experiment, which demonstrates that malignant melanoma skin cancers are mainly caused by a region of the UV spectrum that is not absorbed by ozone and therefore not affected by changes in the ozone layer. When the EPA is not ignoring this result, it is attacking it on the basis that Setlow experimented with fish and that fish are not people. (Ironically, the EPA expresses no such qualms when using rats to determine the carcinogenicity of chemicals.) In the meantime, the EPA has resisted Congressional requests to revise its cost-benefit analysis backing the Montreal Protocol, which was based on the wholly unjustified assumption of 3 million additional skin cancer deaths.

Conclusion

The bottom line is this: Currently available scientific evidence does not support a ban on the production of chlorofluorocarbons (CFCs or freons), halons, and especially methyl bromide. There certainly is no justification for the accelerated phase-out of CFCs, which was instituted in 1992 on nothing more than a highly questionable and widely criticized NASA press conference. Yet because of the absence of full scientific debate of the evidence, relying instead on unproven theories, we now have an international treaty that will conservatively cost the U.S. economy some $100 billion dollars.

The history of the CFC–ozone depletion issue is rife with examples of the breakdown of scientific integrity: selective use of data, faulty application of statistics, disregard of contrary evidence, and other scientific distortions. The policy before and since the Montreal Protocol has been driven by wild and irresponsible scare stories: EPA's estimate of millions of additional skin cancer deaths, damage to immune systems, blind sheep in Chile, the worldwide disappearance of frogs, plankton death, and the collapse of agriculture and ecosystems.

The latest example of "science by press release" is the scare story about a massive ozone hole, fed to the media in September 1995 by the Geneva-based World Meteorological Organization. "At its present rate of growth [it] might grow to record-breaking size . . .," said Rumen Bojkov, a well-known WMO alarmist. But then again, it might not—according to NASA scientist Paul Newman. Australian meteorologist Paul Lehmann agrees: The hole will change its shape, volume, and size daily as it grows; he concludes that its final size is not predictable by comparing data now with those of a year ago.

These scare stories cannot pass what I call the common-sense test: A projected 10 percent UV increase from a worst-case global ozone depletion is the equivalent of moving just 60 miles closer to the equator, say from Washington, D.C., to Richmond, Virginia. New Yorkers moving to Florida experience a more than 200 percent increase in UV because of the change of latitude. Why aren't they dropping like flies? Mail-order nurseries in the upper midwest ship field-grown plants all over the United States. Why don't these plants die?

Scientists involved in ozone research have known these facts from the beginning, but only a few have acknowledged them publicly.

Statement by Daniel L. Albritton, Ph.D.
Director, Aeronomy Laboratory, Environmental Research Laboratories, NOAA, Boulder, CO

Let me underscore right at the outset that the summary that I'm about to give you is not my own assessment. It is indeed the statement of the vast majority of the active and practicing world's ozone researchers regarding the current state of understanding of ozone depletion based upon their own results and their own laboratories, their field observations, and their atmospheric monitoring and their theoretical modeling.

As part of the advice to world government's [sic] on the ozone layer, this ozone community has prepared a series of such state-of-understanding assessments.

In 1985, they prepared this summary, which was used as input by governments for decisions under the Montreal Protocol in 1987.

In 1989, they updated their ozone understanding for the discussion of governments in the London Amendment in 1990. And in 1991, they updated it further to describe the new findings over the last years. And that was input to the Copenhagen Amendments in 1992.

And now, as you have already cited, the world science community has summarized a current viewpoint on ozone depletion, and its executive summary is the article in the short book that you have as part of your package.

These periodic assessments by the community have been deemed to have very high value. They are, first of all, scientific documents. They're based upon the published extensive scientific literature read by colleagues world-wide.

Therefore, they are a solid basis for decision-making, in contrast to anecdotal statements or privately published viewpoints.

They are pure science. The community makes no policy recommendations. That's the job of others, like yourselves, who are entrusted with the public welfare.

Secondly, these are majority statements. In fact, the very, very vast majority. This assessment was prepared by 250 scientists world-wide and peer-reviewed by 150 others.

It's therefore a touchstone of the opinion of the large community. This is in contrast to the sporadic and separate statements reflecting the opinions of either one person or a small group of individuals.

Fourthly, it's an international assessment and it draws from the world scientific community—all nations, all viewpoints, and therefore, international problems can be addressed on a common playing field.

And finally, the scientific scope is comprehensive. Both the natural changes in ozone and the human-induced changes in ozone are considered together. And that's much more comprehensive than a single statement about a single observation or a single publication.

Let me indicate to you the four key conclusions from this.

Albritton then illustrates his main points by referring to a series of charts prepared for the testimony. Only his four main points are listed below. His charts are not included in this book.

The first point is that very large seasonal depletions of the ozone layer continue year after year to be observed in Antarctica. Forty years of Antarctic ozone data records show that this began in the 1970s and has grown larger since then. . . .

The second point I wanted to underscore visually with you is that ozone depletion continues to be observed by the eye over much of the globe.

The second chart shows how the ozone levels have changed over the past 30 years of observations from the ground-based network. . . .

The third point that I wanted to underscore with you is that when ozone is depleted above, ultraviolet radiation increases at the surface.

The third chart shows data taken over long time periods that indicate that any time ozone goes down, . . . that ultraviolet radiation goes up. . . .

Point number four. The maximum ozone losses will likely occur in the next ten years, and thereafter our ozone layer will slowly recover. . . .

In summary, Mr. Chairman, let me just note that this hearing actually began about 20 years ago, when scientists recognized the possibility that our own actions could inadvertently affect the ozone layer. And over that period, some of the world's brightest and most productive atmospheric scientists have sharpened the picture of that initial point. . . .

And so I conclude by noting that while I am speaking for [scientists from the National Academy of Sciences], it is the world-wide ozone research community that you just heard from.

FOOD ADDITIVES

Synthetic chemicals play an important role in the foods humans and animals eat today. They are added to foods for a variety of reasons: to enhance color, odor, and flavor; to prevent clumping in containers; to maintain an emulsified state; to improve nutritional quality; and, perhaps most often, to extend shelf life. A great deal of controversy, however, has surrounded the use of such chemical food additives. In some cases, food additives have been shown to be carcinogenic, teratogenic, mutagenic, or harmful to animals in other ways.

The problem is that the potential health hazard of a food additive can be determined only by experimenting with small animals, such as rats and guinea pigs, and not with humans themselves. How can the results of such tests be extrapolated to humans?

In 1958, the U.S. Congress answered that question—at least for a period of years—by adopting the so-called Delaney Clause as an amendment to the Federal Food, Drug and Cosmetic Act (FFDCA). The Delaney

Clause opened with the now-memorable phrase that "No additive shall be deemed safe if it is found to induce cancer when ingested by man." In practice, the Delaney Clause banned any food additive that had been shown to be carcinogenic *in experimental animals.*

Over time, a number of issues developed regarding the Delaney Clause. For one thing, chemical technology improved, and concentrations of potentially harmful substances could be detected at levels much lower than those at which they were thought to be carcinogenic or otherwise harmful. Also, the standards used for banning or allowing food additives differed for processed foods and for raw foods. In the latter case, the question arose as to what level of pesticide residues should be allowed to remain on unprocessed foods, especially in comparison to the levels of food additives used in processed foods and known to be harmful to human health.

In the mid-1990s, Congress once more wrestled with the question of the Delaney Clause. Some opposing views on this issue were presented in 1996 during hearings on Senate bill 1166, the Food Quality Protection Act. Excerpts reprinted below from those hearings summarize some of the issues involved in writing legislation that will protect human health from potentially harmful food additives.

Reference: "S.1166—Food Quality Protection Act," Hearing before the Committee on Agriculture, Nutrition, and Forestry of the United States Senate, 104th Congress, Second Session, 12 June 1996.

Statement by Bruce Alberts, Ph.D.
President, National Academy of Sciences, and Chair, National Research Council (NRC)

Dr. Alberts's presentation summarized a number of studies conducted under the auspices of the NRC and presented some recommendations about food safety legislation based on the results of those studies.

In 1985, the EPA commissioned the NRC to compare alternative strategies for regulating pesticide residues for their impact on public health and agriculture.
. . .

As your committee knows, the real focus of the study was the fact that EPA is mandated to apply two different standards for the protection of public health from pesticides. One is for raw food and the other is for foods that are processed before sale. In the case of raw food, the EPA must set tolerances for pesticides that protect the public health and that recognize that pesticides confer benefits for an "adequate, wholesome, and economical food supply." In the case of processed foods, considerations of benefits are not allowed; under the Delaney Clause, any additive (including pesticides) that is capable of inducing cancer in humans or animals and becomes concentrated during processing, is prohibited. . . .

After a long review of NRC study results, Dr. Alberts presents the following recommendations:

Recommendations

I believe that several general conclusions can be drawn from our work that are of relevance here to the operations of EPA in registering pesticides. First, it is better to get independent engineering and scientific advice in the planning stages of a program or regulatory action than after implementation has begun. Second, EPA must find a way to give a science and engineering review the same weight as the legal review that proposed regulatory actions and re-search programs receive; a science adviser should be required to either attest to the use of the best available scientific and technical information or explain why it was not used. Third, Mr. Chairman, you and your Colleagues must find ways to encourage EPA to make science and engineering the foundation of EPA actions. An implication of these conclusions is that you must fund research and scientific review activities at a level that will ensure that they are properly complete *before* the implementation of major regulatory or program-matic actions. This will save those who now must pay the bill for unscientific and unfounded actions a considerable amount of money. It will also increase the respect that Americans have for regulatory policy and will reduce more efficiently the threats to their health and the environment.

Specific recommendations of relevance to your pending legislation are:

(1)   Adopt a negligible-risk standard for setting pesticide-residue tolerance limits in all foods (both raw and processed). This could eliminate almost all the potential dietary-related cancer risks posed by these residues.

(2)   In determining tolerances, give special consideration to the toxicologi-cal implications of diets of infants and children.

(3)   Pursue an alternative pest-management approach that takes advan-tage of ecological processes—that is, Ecologically Based Pest Management. A medley of strategies—traditional practices such as the use of crop rotation, beneficial organisms, genetically engineered crop varieties, and narrow-spec-trum pesticides designed to target specific pests—should be considered as options to broad-spectrum pesticide use.

Statement by Albert H. Meyerhoff, Senior Attorney
Natural Resources Defense Council (NRDC)

The NRDC is a non-profit environmental organization. Mr. Meyerhoff opens with a general review of the current state of food additive laws and regulations. He then goes on to speak explicitly about why he believes the Delaney Clause should not be repealed.

The Delaney Clause Should not Be Repealed

It has been the consistent position of the NRDC for several years that the Delaney Clause should be tinkered with, if at all, only if by doing so the number of cancers in American society would be reduced and the public health better protected overall.

Thus, throughout the Bush Administration, the NRDC opposed legislation that would have replaced Delaney with a "cost-benefit" standard for pesticides in the food supply. We were also critical of the initial proposals by the Clinton administration that, while perhaps well intended, in our view would have weakened public health protections. *[citation omitted]* Our position reflects a simple premise: that while no panacea, the famous Clause is grounded in a policy of prevention that remains as sound today as it did when Section 409 was enacted. All reasonable steps should be taken to avoid toxic chemicals in the food supply to which more than two hundred million Americas are exposed daily.

The reasons for this policy remain valid. We *still* do not know whether humans are more or less sensitive than laboratory animals to carcinogens and whether one carcinogen may increase the cancer-causing effects of another. We *still* do not know the cumulative impact of dozens of carcinogens permitted in the food supply and the environment. Our existing tolerance-setting system is entirely predicated on a chemical-by-chemical, crop-by-crop, risk-by-risk approach, grounded in myopia, "managing" cancer, rather than preventing it.

The reality of life is that we are exposed to a multiplicity of toxic substances. Calculating the combined risks of these exposures is problematic at best; some 300 pesticide active ingredients are used on food as well as an imperfectly examined large number of "inert" ingredients. For the most part, existing EPA pesticide tolerances for allowable pesticide residue levels do not even attempt to calculate the aggregate human health risks presented, nor do they address the cumulative and synergistic effects on multiple pathways of exposure. . . .

Delaney is no panacea. It applies only to cancer and then only to certain pesticides that concentrate in certain processed foods. Other carcinogens are subject of FIFRA's cost-benefit standard, as are other pesticide adverse effects, such as their impact on human reproduction, the endocrine system, the immune system, and the neurological system. A virtual mountain of reports from the GAO, NAS, House and Senate committees and others have repeatedly found that our existing pesticide laws are, in the words of one, "fundamentally flawed," archaic, anachronistic, and singularly ineffectual in protecting both human health and our natural environment.

For that reason, *comprehensive* pesticide reform—not just tinkering with Delaney—is long overdue. Unfortunately, attempts at discussions with the industry to find common ground on how to reduce pesticide use and exposure—goals which industry claims to favor—have been unsuccessful. Instead, the pesticide industry seems intent on eviscerating the few safeguards contained in existing law, and that has brought us S. 1166. S. 1166 would only make a failed regulatory scheme far worse, eliminating the one serious protection of public health contained in the current statute while, at the same time, preempting the ability of states and localities to take measures to protect their citizenry from these most toxic of substances. For that reason, we are strongly and unqualifiedly opposed to this ill-considered legislation. . . .

Meyerhoff then goes on to describe what he calls "actual reform: less poison, less risk."

Actual Reform: Less Poison, Less Risk

One of the claimed objectives of S. 1166 is "to safeguard infants and children" from the hazards of pesticides in the environment. It would have precisely the opposite result. Both by replacing the Delaney Clause with a cost-benefit standard and weakening an already flawed pesticide statutory scheme generally, America's children would be placed in even greater jeopardy. Indeed, were this bill a commercial product, its producers would be guilty of false advertising. If the authors of S. 1166 are actually serious about seeking to improve protection of children from pesticides in the environment, they need only turn to the National Academy of Sciences for guidance. It was the express recommendation of the National Academy in *Pesticides in the Diets of Infants and Children* that Federal pesticide laws expressly establish tolerances at levels accounting for the unique effects these substances may have on infants and children, protect against cumulative exposure [and] nondietary intake of pesticides, and eliminate the consideration of benefits. While lip service is paid, none of these recommendations is actually adopted in S. 1166. . . .

Pesticide Reduction: The Pollution Prevention Solution

If the Delaney Clause is to be replaced with a new policy on carcinogens, its linchpin must be (1) to greatly improve the public's "right to know" they are being exposed to pesticides; and (2) to reduce overall hazards caused by pesticides by eliminating them at the source.

Right to Know.—Probably the greatest success story in environmental protection (with the possible exception of the removal of lead from gasoline) has been improving the public's right to know. For example, the Toxic Release Inventory (TRI), enacted by Congress after the tragedy in Bhopal, now requires public notice of exposure to several hundred known hazardous substances when emitted by the petroleum and chemical industries. Attempts to weaken TRI, both here in Congress and the courts, have failed. And rather than notify workers or citizens living near chemical plants, industry has frequently dramatically reduced emissions—often saving money by doing so—a true "win-win". Similarly, California's experiment with Proposition 65—to be preempted by S. 1166—has been equally successful. Adopted by initiative in 1986, the act placed the burden of proof on polluters to establish the safety of toxic exposures, with no consideration of benefits, or a cancer warning is mandatory. Rather than provide the required "clear and reasonable" warnings, numerous consumer products—from drinking water faucets to paint strippers—have been reformulated.

There is a lesson here. The best means to greatly reduce exposure to carcinogens—the goal of Delaney—may well be simply to *tell the people* when such substances are present in food, in home use products, and in the fields. The marketplace and freedom of choice will do the rest.

Use Reduction.—Unfortunately, while many other industries have engaged in good faith efforts to reduce emissions of toxic substances into the environment, agriculture and the pesticide industry have woefully failed to do so. Yet, numerous reports document the potential and importance of reducing the overall use of pesticides. According to the *National Academy of Science's Soil and Water Quality: An Agenda for Agriculture* NAS report:

Source control to reduce the total mass of pesticides applied to cropping systems should be the fundamental approach to reducing pesticide losses from farming systems.

Several European nations, including Sweden, Denmark, and the Netherlands have adopted national programs that incorporate the fundamental approach of pesticide use reduction. Concern about environmental pollution has prompted these countries to initiate programs aimed at reducing the use and emissions of, and dependence on, pesticides while maintaining viable levels of crop protection without decreasing crop yields.

Numerous methods are available to reduce agriculture's use of and reliance on pesticides. The National Academy of Sciences in their report, *Alternative Agriculture*, documented the potential for reducing pesticide use through the adoption of integrated pest management and other practices and systems for agricultural sustainability. Such practices can lower costs for farmers and pest managers and in many cases increase the quality, productivity, and yields.

Statement by Dean Kleckner, President
American Farm Bureau Federation, and John Cady, President and Chief Executive Officer, National Food Processors Association (NFPA)

Here is industry's side of the debate. Kleckner, head of the nation's largest organization of farmers and ranchers, spoke first.

It has been said that legislation is driven by crisis. Based on that principle, reform of the Delaney Clause should become a reality very soon. The long-awaited Delaney disaster has begun for U.S. farmers. Today, the Delaney Clause is forcing the unjustified cancellation of safe and effective crop production products used on a broad range of fruits, vegetables, and major agricultural commodities.

From California grapes to Indiana soybeans, Kansas wheat, Alabama peanuts, Florida tomatoes and apples in Vermont, farmers in virtually every state are the victims of these unwarranted cancellations. All are caught in the Delaney net. Some of these farmers may not be able to continue commercial production of some crops. Farmers and consumers can no longer wait for reform of the Delaney Clause. We commend you, Mr. Chairman and other Members of this committee, for your leadership and support for reforming the Delaney Clause. We sincerely thank you for holding this hearing today. . . .

Kleckner then provides a brief discussion of the history of the Delaney Clause and its current status in agriculture.

Background

The goal of the Delaney Clause is admirable—to prevent cancer-causing agents from entering the food supply. No one argued with that goal then and no one would argue the merits of the goal today. Except the science on which the Delaney Clause was based upon then—the best available science—is no longer the best available science. Incredible improvements in scientific method, knowledge, and technology have since rendered the Delaney Clause obsolete.

Strict enforcement of Delaney's unreasonable zero risk standard will force the loss of crop protection products merely because they are detectable, not because they are unsafe. Delaney's rigid zero tolerance standard means simply this: If a laboratory animal develops tumors after being continually fed the maximum dose they can tolerate, Delaney tells us that any dose above zero of the same substance will cause cancer in humans. That same Delaney logic applied elsewhere illustrates the folly of current policy. For example, if a child becomes ill after consuming two pounds of potato chips, Delaney tells us potato chips in any amount create a certain hazard to health.

New scientific evidence adds further fuel to the public firestorm clamoring for the repeal of the Delaney Clause. Earlier this year, the National Research Council (NRC) released the report, "Carcinogens and Anticarcinogens in the Human Diet." The charge to the researchers who developed the report was to "examine the occurrence . . . and potential role of natural carcinogens in the causation of cancer (in humans), including relative risk comparison with synthetic carcinogens and a consideration of anticarcinogens." The reports draws several conclusions, which include:

— "The great majority of individual naturally occurring and synthetic food chemicals are present in the human diet at levels so low that they are unlikely to pose an appreciable cancer risk."

— "Most naturally occurring minor dietary constituents occur at levels so low that any biologic effect, positive or negative, is unlikely. The synthetic chemicals in our diet are far less numerous than the natural and have been more thoroughly studied, monitored, and regulated. Their potential biologic effect is lower."

In other words, natural chemicals present in the foods we eat pose a greater cancer risk than synthetic substances. However, both natural and synthetic substances are present in our diets "at levels so low that they are unlikely to pose an appreciable cancer risk." . . .

Mr. Cady then follows with remarks about the impact of the Delaney Clause on the processed-foods industry.

Impact of the Delaney Clause on the Processed-Foods Industry

NFPA has led the food industry's efforts to reform the pesticide regulatory process, as well as to resolve the Delaney paradox for the past 10 years. NFPA has supported legislation that promotes sound scientific judgment in pesticide tolerances decisions, assures that tolerance decisions are based on accurate exposure data, requires renewal of pesticide tolerances to assure compliance with current safety standards, facilitates minor use registrations and provides for national uniformity of pesticide tolerances.

Consistent with these objectives, NFPA strongly supports S. 1166, the Lugar-Pryor "Food Quality Protection Act of 1995." S. 1166 would make important improvements in both the Federal Insecticide, Fungicide, and Rodenticide Act (FIFRA) and the Federal Food, Drug, and Cosmetic Act (FD&C Act). It would establish a consistent negligible risk standard for pesticide tolerances for raw and processed food, assume appropriate consideration of pesticide benefits, and provide for national uniformity for tolerances meeting current safety standards. Moreover, S. 1166 contains specific provisions, which we strongly support, that would require EPA to implement procedures that ensure that pesticide tolerances adequately safeguard the health of infants and children and that would facilitate the registration of minor use pesticides. . . .

Negligible Risk Standard

Under current law, there are two different legal standards for pesticide residues in food. Tolerances for pesticide residues in raw agricultural commodities are issued under Section 408 of the FD&C Act, which contains a general safety standard, no anticancer clause and requires consideration of benefits. Tolerances for pesticide residues that concentrate in, or are applied to, processed foods are issued under Section 409 of the FD&C Act, which contains the Delaney anti-cancer clause and does not expressly provide for consideration of benefits.

In 1987, the NAS recommended that pesticide residues in raw and processed food should be regulated on the basis of a single negligible risk standard. Consistent with the NAS recommendation, S. 1166 would eliminate application of section 409 to pesticide residues in processed food and would provide that all pesticide residues in food are governed by the tolerance provisions of section 408, which governs raw food commodities.

S. 1166 would require EPA to set a tolerance at a level adequate to protect the public health taking into account, amcng other relevant factors, the validity, completeness and reliability of available data from studies of the pesticide chemical, the nature of any toxic effects caused by the chemical, available information, and reasonable assumptions concerning the relationship of the data to human risk and to the dietary exposure levels of consumers.

A tolerance level for a pesticide residue in food would be deemed to be adequate to protect the public health if the dietary risk to consumers from exposure to a pesticide is negligible. This would implement the NAS recommendation for a uniform negligible risk standard for pesticide residues in foods, would give EPA flexibility to ignore *de minimis* or insignificant risks and would permit the Agency to focus its limited resources on the highest risk pesticides.

The definition of "negligible risk" would not identify a specific level of risk that would be considered negligible or a numerical expression of that level, because science and the degree of knowledge and confidence in cancer risk assessment is constantly evolving and improving, and EPA needs to preserve the ability to keep pace with the developing science of risk assessment. The Administrator would be required, however, to set forth by regulation the factors and methods, including tests which are appropriate for the determination of dietary risk and most likely dietary exposure, for the determination of negligible dietary risk.

EPA would be required, where reliable data are available, to calculate dietary exposure on the basis of the percent of food actually treated with a pesticide and the actual residue levels of the pesticide that occur in food. This would help avoid today's exaggerated and unjustified exposure assumptions and would assist in developing more realistic risk projections.

Benefits

S. 1166 would provide that EPA may establish a tolerance for a pesticide residue posing greater than a negligible risk, if the Agency determines that the risk is reasonable because (1) the risk to human health from loss of the pesticide outweighs the dietary risk attributable to the pesticide, (2) likely alternative pesticides or pest control methods pose greater risks to the public than the dietary risks attributable to the pesticide, or (3) loss of the pesticide would limit the availability of an adequate, wholesome, or economical food supply, taking into account regional and national effects. This would assure that pesticide tolerance decisions are not made in isolation, that EPA fully considers all relevant factors and that a complete risk-benefit evaluation is conducted. In this manner, EPA will be able to set reasonable priorities and direct its limited resources against the most hazardous and least beneficial pesticides.

Food use pesticides provide important benefits for American agriculture. Loss of pesticides can cause reductions in crop yield and farm income and decreases in agricultural employment. Pesticides also play an essential role in providing a wholesome, healthy, nutritious, and affordable food supply. The Department of Health and Human Services (HHS), the Surgeon General, other Government agencies, and prominent scientific and medical organizations have increasingly stressed the value of a nutritious diet, including ample amounts of fresh fruits and vegetables, in maintaining health and preventing disease. The loss of important food use pesticides can result in heightened risk of disease causing organisms in food, food price increase, adverse effects on food quality, and reduction in the availability of nutritional food choices, particularly for low income consumers. In view of the important benefits of food use pesticides, we believe that it is essential that EPA be permitted to retain the authority to balance benefits against risks in pesticide tolerance decisions, especially since the Agency's risk estimates are based on multiple conservative assumptions and extrapolations.

CLIMATE CHANGE

Issues such as global warming (climate change) and ozone depletion present society with difficult choices. In such issues, scientific evidence may be difficult to obtain, may be needed over relatively long periods of time, and may present profound challenges in terms of interpretation and extrapolation. Governments are hard-pressed to develop long-term "solutions" for problems that may not be clearly defined or whose effects may be debatable.

Yet, at the end of the 1990s, world governments appear largely to have accepted the reality of global warming and ozone depletion and have started to draft agreements that will help reduce the severity of these problems and their long-term effects on human societies.

In December 1997, representatives of nations around the world met to draft an agreement about goals and methods for reducing the amount of global warming created by human activities. Major portions of the Kyoto Protocol on climate change are provided below. The complete treaty can be found on the Internet at http://www.unfccc.de/.

Kyoto Protocol to the United Nations Framework Convention on Climate Change

The Parties to the Protocol,
    Being Parties to the United Nations Framework Convention on Climate Change, hereinafter referred to as "the Convention",
    In pursuit of the ultimate objective of the Convention as stated in its Article 2,
    Recalling the provisions of the Convention,
    Being guided by Article 3 of the Convention,
    Pursuant to the Berlin Mandate adopted by decision 1/CP.1 of the Conference of the Parties to the Convention at its first session,
    Have agreed as follows:

Article 1

Article 1 defines seven fundamental terms used in the Convention.

Article 2

1. Each Party included in Annex I in achieving its quantified emission limitation and reduction commitments under Article 3, in order to promote sustainable development, shall:
    (a) Implement and/or further elaborate policies and measures in accordance with its national circumstances, such as:

     (i)   Enhancement of energy efficiency in relevant sectors of the national economy;

    (ii)   Protection and enhancement of sinks and reservoirs of greenhouse gases not controlled by the Montreal Protocol, taking into account its commitments under relevant international environmental agreements; promotion of sustainable forest management practices, afforestation and reforestation.

   (iii)   Promotion of sustainable forms of agriculture in light of climate change considerations;

   (iv)   Promotion, research, development and increased use of new and renewable forms of energy, of carbon dioxide sequestrian technologies and of advanced and innovative environmentally sound technologies:

    (v)   Progressive reduction or phasing out of market imperfections, fiscal incentives, tax and duty exemptions and subsidies in all greenhouse gas emitting sectors that run counter to the objective of the Convention and apply market instruments;

   (vi)   Encouragement of appropriate reforms in relevant sectors aimed at promoting policies and measures which limit or reduce emissions of greenhouse gases not controlled by the Montreal Protocol;

   (vii)   Measures to limit and/or reduce emissions of greenhouse gases not controlled by the Montreal Protocol in the transport sector.

  (viii) Limitation and/or reduction of methane through recovery and use in waste management, as well as in the production, transport and distribution of energy.

  (b) Cooperate with other such Parties to enhance the individual and combined effectiveness of their policies and measures adopted under this Article, pursuant to Article 4, paragraph 2(e)(i), of the Convention. To this end, these Parties shall take steps to share their experience and exchange information on such policies and measures, including developing ways of improving their comparability, transparency and effectiveness. The Conference of the Parties serving as the meeting of the Parties to this Protocol shall, at its first session or as soon as practicable thereafter, consider ways to facilitate such cooperation, taking into account all relevant information.

2. The Parties included in Annex I shall pursue limitation or reduction of emissions of greenhouse gases not controlled by the Montreal Protocol from aviation and marine bunker fuels, working through the International Civil Aviation Organization and the International Maritime Organization, respectively.

3. The Parties included in Annex I shall strive to implement policies and measures under this Article in such a way as to minimize adverse effects, including the adverse effects of climate change, effects on international trade, and social, environmental and economic impacts on other Parties, especially developing country Parties and in particular those identified in Article 4, paragraph 8 and 9 of the Convention, taking into account Article 3 of the Convention. The Conference of the Parties serving as the meeting of the Parties to this Protocol may take further action, as appropriate, to promote the implementation of the provisions of this paragraph.

4. The Conference of the Parties serving as the meeting of the Parties to this Protocol, if it decides that it would be beneficial to coordinate any of the policies and measures in paragraph 1(a) above, taking into account different national circumstances and potential effects, shall consider ways and means to elaborate the coordination of such policies and measures.

Article 3

1. The Parties included in Annex I shall, individual or jointly, ensure that their aggregate anthropogenic carbon dioxide equivalent emissions of the greenhouse gases listed in Annex A do not exceed their assigned amounts, calculated pursuant to their quantified emission limitation and reduction commitments inscribed in Annex B and in accordance with the provisions of this Article, with a view to reducing their overall emissions of such gases by at least 5 per cent below 1990 levels in the commitment periods 2008 to 2012.

2. Each Party included in Annex I shall, by 2005, have made demonstrable progress in achieving its commitments under this Protocol.

3. The net changes in greenhouse gas emissions from sources and removals by sinks resulting from direct human-induced land use change and forestry activities, limited to afforestation, reforestation, and deforestation since 1990, measured as verifiable changes in stocks in each commitment period shall be used to meet the commitments in the Article of each Party included in Annex I. The greenhouse gas emissions from sources and removals by sinks associated with those activities shall be reported in a transparent and verifiable manner and reviewed in accordance with Articles 7 and 8.

4. Prior to the first session of the Conference of the Parties serving as the meeting of the Parties to this Protocol, each Party included in Annex I shall provide for consideration by the Subsidiary Body for Scientific and Technological Advice data to establish its level of carbon stocks in 1990 and to enable an estimate to be made of its changes in carbon stocks in subsequent years. The Conference of the Parties serving as the meeting of the Parties to this Protocol shall, at its first session or as soon as practicable thereafter, decide upon modalities, rules and guidelines as to how and which additional human-induced activities

related to changes in greenhouse gas emissions and removals in the agricultural soil and land use change and forestry categories, shall be added to, or subtracted from, the assigned amount for parties included in Annex I, taking into account uncertainties, transparency in reporting, verifiability, the methodological work of the Intergovernmental Panel on Climate Change, the advice provided by the Subsidiary Body for Scientific and Technological Advice in accordance with Article 5 and the decisions of the Conference of the Parties. Such a decision shall apply in the second and subsequent commitment periods. A Party may choose to apply such a decision on these additional human-induced activities for its first commitment period, provided that these activities have taken place since 1990.

5. The Parties included in Annex I undergoing the process of transition to a market economy whose base year or period was established pursuant to decision 9/CP.2 of the Conference of the Parties at its second session, shall use that base year or period for the implementation of their commitments under this Article. Any other Party included in Annex I undergoing the process of transition to a market economy which has not yet submitted its first national communication under Article 12 of the Convention may also notify the Conference of the Parties serving as the meeting of the Parties to this Protocol that it intends to use a historical base year or period other than 1990 for the implementation of its commitments under this Article. The Conference of the Parties serving as the meeting of the Parties to this Protocol shall decide on the acceptance of such notification.

6. Taking into account Article 4, paragraph 6, of the Convention, in the implementation of their commitments under this Protocol other than those in this Article, a certain degree of flexibility shall be allowed by the Conference of the Parties serving as the meeting of the Parties to this Protocol to the Parties included in Annex I undergoing the process of transition to a market economy.

7. In the first quantified emission limitation and reduction commitment period, from 2008 to 2012, the assigned amount for each party included in Annex I shall be equal to the percentage inscribed for it in Annex B of its aggregate anthropogenic carbon dioxide equivalent emissions of the greenhouse gases listed in Annex A in 1990, or the base year or period determined in accordance with paragraph 5 above, multiplied by five. Those Parties included in Annex I for whom land use change and forestry constituted a net source of greenhouse gas emissions in 1990 shall include in their 1990 emissions base year or period the aggregate anthropogenic carbon dioxide equivalent emissions minus removals in 1990 from land use change for the purposes of calculating their assigned amount.

8. Any Party included in Annex I may use 1995 as its base year for hydrofluorocarbons, perfluorocarbons and sulphur hexafluoride, for the purposes of the calculation referred to in paragraph 7 above.

9. Commitments for subsequent periods for Parties included in Annex I shall be established in amendments to Annex B to this Protocol, which shall be adopted in accordance with the provisions of Article 20, paragraph 7. The Conference of the Parties serving as the meeting of the Parties to this Protocol shall initiate the consideration of such commitments at least seven years before the end of the first commitment period mentioned in paragraph 7 above.

10.   Any emission reduction units, or any part of an assigned amount, which a Party acquires from another Party in accordance with the provisions of Article 6 and of Article 16 bis shall be subtracted from the assigned amount for that Party.

12.   Any certified emission reductions which a Party acquires from another Party in accordance with the provisions of Article 12 shall be added to the assigned amount for that Party.

13.   If the emissions of a Party included in Annex I during a commitment period are less than its assigned amount under this Article, this difference shall, on request of that Party, be added to the assigned amount for that Party for subsequent periods.

14.   Each Party included in Annex I shall strive to implement the commitments mentioned in paragraph 1 above in such a way as to minimize adverse social, environmental, and economic impacts on developing country Parties, particularly those identified in Article 4, paragraphs 8 and 9, of the Convention. In line with relevant decisions of the Conference of the Parties on the implementation of those paragraphs, the Conference of the Parties serving as the meeting of the Parties to this Protocol shall, at its first session, consider what actions are necessary to minimize the adverse effects of climate change and/or the impacts of response measures on Parties referred to in those paragraphs. Among the issues to be considered shall be the establishment of funding, insurance, and transfer of technology.

Article 4

This article outlines the responsibilities of Parties that "agree to jointly fulfill their commitments," such as regional groups of countries. Examples include the following.

1. Any Parties included in Annex I that have agreed to jointly fulfil their commitments under Article 3 shall be deemed to have met those commitments provided that their total combined aggregate anthropogenic carbon dioxide equivalent emissions of the greenhouse gases listed in Annex A do not exceed their assigned amounts calculated pursuant to their quantified emission limitation and reduction commitments inscribed in Annex B and in accordance with provisions of Article 3. . . .

Article 5

Article 5 discusses methods by which Parties are to estimate the anthropogenic emission by sources and removals by sinks of all greenhouse gases covered by the Convention.

Article 6

1. For the purpose of meeting its commitments under Article 3, any Party included in Annex I may transfer to, or acquire from, any other such Party emission reduction units resulting from projects aimed at reducing anthropogenic emissions by sources or enhancing anthropogenic removals by sinks of greenhouse gases in any sector of the economy, provided that:

The restrictions governing such trade offs are then discussed.

Article 7

Article 7 outlines the methods by which Parties are to collect, collate and report data on greenhouse gas emissions.

Article 8

This article defines the system by which data submitted by the Parties is to be reviewed and evaluated. Its main features include the following.

1. The information submitted under Article 7 by each party included in Annex I shall be reviewed by expert review teams pursuant to the relevant decisions of the Conference of the Parties . . .
2. Expert review teams shall be coordinated by the secretariat and shall be composed of experts selected from those nominated by Parties to the Convention . . .
3. The review process shall provide a thorough and comprehensive technical assessment of all aspects of the implementation by a Party of this Protocol. . . .

Article 9

The ongoing group known as the Conference of the Parties serving as the meeting of the Parties to this Protocol is defined and its duties listed in this article.

Article 10

All Parties, taking into account their common but differentiated responsibilities and their specific national and regional development priorities, objectives and circumstances, without introducing any new commitments for Parties not included in Annex I, but reaffirming existing commitments in Article 4, paragraph 1, of the Convention, and continuing to advance the implementation of these commitments in order to achieve sustainable development, taking into account Article 4, paragraphs 3, 5 and 7, of the Convention, shall:

(a) Formulate, where relevant and to the extent possible, cost-effective national, and where appropriate regional programmes to improve the quality of local emission factors, activity data and/or models which reflect the socio-economic conditions of each Party for the preparation and periodic updating of national inventories of anthropogenic emissions by sources and removals by sinks of all greenhouse gases not controlled by the Montreal Protocol, using comparable methodologies to be agreed upon the Conference of the Parties, and consistent with the guidelines for national communications adopted by the Conference of the Parties;

(b) Formulate, implement, publish and regularly update national and, where appropriate, regional programmes containing measures to mitigate climate change and measures to facilitate adequate adaptation to climate change: . . .

(c) Cooperate in the promotion of effective modalities for the development, application and diffusion of, and take all practicable steps to promote, facilitate and finance, as appropriate, the transfer of, or access to, environmentally sound technologies, know-how, practices and processes pertinent to climate change, in particular to developing countries . . .

(d) Cooperate in scientific and technical research and promote the maintenance and the development of systematic observation systems and development of data archives to reduce uncertainties related to the climate system, the adverse impacts of climate change and the economic and social consequences of various response strategies, and promote the development and strengthening of endogenous capacities and capabilities to participate in international and intergovernmental efforts, programmes and networks on research and systematic observation, taking into account Article 5 of the Convention.

(e) Cooperate in and promote at the international level, and, where appropriate, using existing bodies, the development and implementation of education and training programmes, including the strengthening of national capacity building, in particular human and institutional capacities and the exchange or secondment of personnel to train experts in this field, in particular for developing countries, and facilitate at the national level public awareness and public access to information on climate change. . . .

Article 11
Some of the financial issues involved in carrying out the conditions of the Convention are described in this article.

Article 12
This article defines the concept of a "clean development mechanism," shows how it applies to both Parties in Annex I and those not in Annex I, and describe how the mechanism can be used to reduce greenhouse gas emissions.

Article 13
The agency known as The Conference of the Parties serving as the meeting of the Parties to this Protocol is defined, and its duties and responsibilities are outlined and explained.

Article 14
This article establishes the secretariat of the Convention and outlines its functions.

Article 15
The duties of the Subsidiary Body for Scientific and Technological Advice and the Subsidiary Body for Implementation of the Convention are defined in this article.

Articles 16, 16 bis, and 17
Additional responsibilities of the Conference of the Parties serving as the meeting of the Parties to this Protocol are defined in these three short articles.

Article 18
The provisions of Article 14 of the Convention on settlement of disputes shall apply mutatis mutandis to this Protocol.

Article 19
Article 19 describes the process by which amendments to the Convention may be made.

Article 20
This article deals with existing annexes to the Protocol and annexes that might be added in the future. It explains how they are to be adopted and implemented.

Article 21
1. Each Party shall have one vote, except as provided for in paragraph 2 below.
2. Regional economic integration organizations, in matters within their competence, shall exercise their right to vote with a number of votes equal to the number of their member States which are Parties to this Protocol. Such an organization shall not exercise its right to vote if any of its member States exercises its right, and vice versa.

Article 22
The Secretary-General of the United Nations shall be the Depositary of this Protocol.

Article 23
Provisions for the signing of the Protocol are outlined.

1. This Protocol shall be open for signature and subject to ratification, acceptance or approval by States and regional economic integration organizations which are Parties to the Convention. It shall be open for signature at United Nations Headquarters in New York from 16 March 1998 to 15 March 1999. . . .

Article 24
1. This Protocol shall enter into force on the ninetieth day after the date on which not less than 55 Parties to the Convention, incorporating Parties included in Annex I which accounted in total for at least 55 percent of the total carbon dioxide emissions for 1990 of the Parties

included in Annex I, have deposited their instruments of ratification, acceptance, approval or accession. . . .

Article 25

This article states that there are no issues left unresolved within this protocol.

Article 26

1. At any time after three years from the date on which this Protocol has entered into force for a Party, that Party may withdraw from this Protocol by giving written notification to the Depositary. . . .

Article 27

The original of this Protocol, of which the Arabic, Chinese, English, French, Russian, and Spanish texts are equally authentic, shall be deposited with the Secretary-General of the United Nations.
Done at Kyoto this tenth day of December one thousand nine hundred and ninety-seven.

Annex A

Greenhouse gases
Carbon dioxide (CO_2)
Methane (CH_4)
Nitrous oxide (N_2O)
Hydrofluorocarbons (HFCs)
Perfluorocarbons (PFCs)
Sulphur hexafluoride (SF_6)

Sectors/source categories

Energy
    Fuel consumption
        Energy industries
        Manufacturing industries and construction
        Transport
        Other sectors
        Other
    Fugitive emissions from fuels
        Solid fuels
        Oil and natural gas
        Other

Industrial processes
    Mineral processes
    Chemical industry
    Metal production
    Other production
    Production of halocarbons and sulphur hexafluoride

Consumption of halocarbons and sulphur hexafluoride
Other

Solvent and other product use

Agriculture
    Enteric fermentation
    Manure management
    Rice cultivation
    Agricultural soils
    Prescribed burning of savannas
    Field burning of agricultural residues
    Other

Waste
    Solid waste disposal on land
    Wastewater handling
    Waste incineration
    Other

Annex B

Party	Quantified emission limitation or reduction commitment (percentage of base year or period)
Australia	108
Austria	92
Belgium	92
Bulgaria*	92
Canada	94
Croatia*	95
Czech Republic*	92
Denmark	92
Estonia*	92
European Community	92
Finland	92
France	92
Germany	92
Greece	92
Hungary*	94
Iceland	110
Ireland	92
Italy	92
Japan	94
Latvia*	92
Liechtenstein	92
Lithuania*	92
Luxembourg	92

Monaco	92
Netherlands	92
New Zealand	100
Norway	101
Poland*	94
Portugal	92
Romania*	92
Russian Federation*	100
Slovakia*	92
Slovenia*	92
Spain	92
Sweden	92
Switzerland	92
Ukraine*	100
United Kingdom	92
United States of America	93

*Countries that are undergoing the process of transition to a market economy.

CHEMICAL WEAPONS CONVENTION

The nations of the world have been trying for almost a century to find a way to ban chemical and biological weapons. Various treaties have been drafted and signed, but none has provided the assurance that these deadly weapons will not be used by a combatant in an international or regional conflict. The most recent attempt to ban the use of chemical and biological weapons was the Chemical Weapons Convention (CWC). The CWC was negotiated in Geneva and the final draft signed by 65 nations on 3 September 1991. For more than five years, then, efforts were made to obtain ratification of the treaty in the United States Senate. Over the period 1994 to 1996, a number of hearings were held in the Senate on the CWC. At these hearings, arguments both for and against approval of the treaty were presented. The two selections below illustrate the kinds of points made by proponents and opponents of the treaty.

Reference: "Chemical Weapons Convention" (Treaty Doc. 103-21), Hearings before the Committee on Foreign Relations of the United States Senate, 103rd Congress, Second Session, 22 March, 13 April, 13 and 17 May, and 9 and 23 June 1994.

Statement by Francis P. Hoeber
former Deputy Under Secretary of the Army

Thank you for the opportunity to testify before you today. As you may recall, I was the Deputy Under Secretary of the U.S. Army during the Reagan Administration, and I led the Army's efforts in chemical warfare matters between 1981 and 1986. I appreciate the chance to raise some of my concerns about the Chemical Warfare Convention and to go on record recommending that the Convention not be ratified.

There are a number of reasons why I believe the treaty should not be ratified that relate to the foreign policy concerns of this Committee. However, given the limited time available to me to offer my comments, I can only summarize a few of my concerns, and those only briefly. I hope, should I succeed in raising any doubts, that the Committee will pursue this matter most carefully and actively seek out more information and insights on the problems of the Convention prior to concluding their deliberations.

First, it should be clearly understood by the Committee that, contrary to the statements of the proponents this treaty won't "rid the world of chemical weapons."

This is true for several reasons:

• The treaty will clearly *not* be global, and, in fact, several countries of foreign policy concern to the United States, namely Libya and North Korea, have announced that they will not sign the Convention.

• Even if the treaty *were* universally signed, there is no assurance that it will be abided by in the case of all the signatory countries and little confidence that violations of military significance in countries other than the major powers would come to the attention of the United States. I don't want to get into the details of the viability and efficacy of the verification regime—burdensome, onerous, and expensive as it is—because others testifying today will do so, but I would like to suggest that your Committee ask the intelligence community to honestly assess whether we could have a reasonable likelihood of detecting a militarily significant violation that was done with an intent to conceal. I do not believe such detection is probable.

Thus, even under the best of circumstances, chemical weapons will remain a threat. This treaty does not change that fact of life. The only chemical weapons which the treaty will eliminate are those which would be eliminated in the absence of a treaty—primarily ours, and, if we fund the process, those declared stocks of the former Soviet Union.

Second, the Chemical Weapons Convention will reduce the ability of the United States to deter chemical warfare by eliminating retaliation in kind from the menu of deterrence options available to a President.

The issue of whether the United States should eliminate the bulk of its chemical weapons stockpile has long since been decided, and correctly in my view. The unitary munitions have been obsolete for some time. However, the decision to eliminate the more modern although minuscule chemical retaliatory capability—the binaries—has been a function of the same wishful thinking that has resulted in this treaty. Even though the use of the retaliatory binaries was forsworn by the prior Administration, in my view their existence *per se* created an uncertainty in the minds of anyone who might be considering using chemical weapons against our forces or against those of one of our allies, and this uncertainty has enhanced deterrence.

This treaty will eliminate the flexibility that some retaliation in kind capability provides to our President, jeopardizing in particular the potential security of U.S. military personnel.

Third, this treaty, by purporting to "rid the world of chemical weapons," guarantees complacency which will have an inevitable effect of reducing the level of effort going into defenses against chemical weapons.

What will suffer will not be the ability to acquire chemical weapons for those who would use them, but rather our ability to provide for the defenses of our own forces.

History has shown that in every case, the ratification of a treaty by the Senate has been seen as reducing the threat to the United States military, whether such a reduction actually occurs or not. As a result, expenditures on research, development, and acquisition of measures to defend against that threat are cut drastically. The only people who benefit by this have been our adversaries.

I cite as an example the Biological Warfare Convention, which resulted in more than a 50% cut in research and development of protection against biological agents within a year after its ratification, and a reduction in our ability to determine the actual facts of the Soviet biological warfare activities—for example, at Sverdlovsk—until the change in the world order put the Russians in a position to discuss the matter. Another example is that of the SALT treaties, which resulted in severe cuts in research and development on both U.S. ballistic missile penetration measures and defense of the U.S. population against third nation and rogue nation ballistic missile threats.

It should be noted, by the way, that funding cuts do not have a linear effect. The cut of 50% in the funding of biological defense efforts resulted in a cut in effectiveness of far more than that amount, perhaps as much as 90%.

I also note that the limitations of research and development efforts on both biological warfare defense and ballistic missile defense have come to be regretted as it has become clearer and clearer that there are countries whose activities are not as constrained by what has come to be referred to as "international norms."

As a former Army official, and as the sometimes-called "mother of the chemical Corps," I have very serious concerns about a reduction in the attention paid to maintaining chemical defense preparedness. It is far more difficult to provide for defenses than it is for an adversary to manufacture offensiver [sic] materiel in this case. Many, including those in Congress, have supported the Department of Defense's efforts to provide American troops with quality chemical protective gear. There is no disputing the fact that these programs had a relatively low priority even in the face of a clear and defined threat from major adversaries during the Cold War. I am particularly concerned about what will happen to these programs should this treaty be approved, for the United States military remains in need of vastly improved protective gear, equipment modifications, decontamination capabilities, and training. One can only imagine what low priority such pursuits will enjoy when there is not supposed to be any threat at all. And history has shown us that adversaries—and there are always unexpected adversaries—will take advantage of such weaknesses.

Fourth, and the final point I wish to make, the Chemical Weapons Convention makes a mockery of arms control objectives themselves and sets an extremely dangerous precedent.

To see what this treaty does to the concepts of foreign relations and international law, it is necessary to understand the context that has driven the completion of the CWC. And that context is that the only reason for this Convention is that an existing Convention—the Geneva Protocol, which prohibits first use of chemical weapons—is inadequate and has failed.

Why has it failed? The Geneva Protocol has failed because there are no international mechanisms for enforcing it and "international norms" have not served to accomplish that enforcement. Nor does it strengthen the "international norms." And it is use, not possession, of such weapons that is the fundamental problem faced by the nations of the world.

Let's look at a specific example. When the Iraqis used chemical weapons against Iran, the international community was horrified. The "international norm" against that use was as strong as it has ever been. Yet the international community was still not able to figure out what to do about it. No sanctions were ever imposed; no resolutions were ever passed condemning that use; nothing was ever done about that treaty violation except to increase the arguments by some that if that treaty had failed, another treaty should be concluded.

The logic of this escapes me. If "international norms" were not strong enough to effectively isolate or punish a country whose violations resulted in the death of a significant number of people, what makes us think that the "international norms" would be strong enough to prevent or punish the violator of a new treaty which involved only the possession of a prohibited weapon? This so-called solution is an exercise in escapism. It cheapens the currency of arms control as well as avoids the real problem.

That real problem is what to do about violators. We clearly haven't solved that problem, either nationally or internationally. Witness the situation in North Korea today relative to the nuclear non-proliferation treaty.

Unless we resolve the broader issue, in my view the Chemical Weapons Convention is worse than meaningless. It's a pretense. This is the foreign policy matter which should be addressed by this Committee.

Statement by Frederick L. Webber

President and CEO, Chemical Manufacturers Association

Mr. Chairman, members of the Senate Foreign Relations Committee. I am Fred Webber, President and CEO of the Chemical Manufacturers Association (CMA). I am here today to reiterate the U.S. Chemical industry's unwavering support for the Chemical Weapons Convention (CWC).

The chemical industry is one of many business sectors that strongly support the CWC. I'm proud to note that the Business Executives for National Security, which represents a broad section of American commerce, and CMA, have formed the Business Alliance to Protect Americans from Chemical Weapons. The business community is united in its support for this Convention.

The members of this Committee may well wonder what would prompt a highly regulated industry like chemical manufacturers to support a treaty like the CWC. After all, it will bring additional regulation to a highly regulated industry. The answer is simple.

- Our commitment to the CWC stems from a shared belief that chemical weapons *must* be banned.
- Our commitment to the CWC is recognition that commercial chemical manufacturers can and must help achieve the goal of a total ban.
- Our commitment to the CWC comes from our responsibility to help prevent legitimate chemical products being used for illegal purposes. That commitment is framed in the Responsible Care® program, our industry's program for the responsible management of chemicals and chemical products.

For our industry, the many benefits of this treaty *far* outweigh the cost. The Convention will not have the catastrophic impact on American business that a few critics claim. Several treaty opponents have stated their belief that 25,000 American companies will be subject to on-site inspections, that American industry's most important business secrets will be compromised, and that America's economic security will be put at risk.

Those arguments are not true; they are not supported by the facts.

We have studied this treaty in great detail; we have put it to the test. We think the CWC is a good deal for American industry.

The far-reaching impact of the Chemical Weapons Convention has been over-stated. Fewer than 2,000 American companies will be directly affected. Of that number, less than 200 are likely to ever have an on-site inspection. The vast majority of the affected facilities will be required to keep records and fill out a yearly one-page report. In our view, that's a manageable burden. In an industry already subject to on-site inspections by the Occupational Safety and Health Administration and the Environmental Protection Agency, and which fills out more forms than almost any other industrial sectors, the considerable benefits of the CWC outweigh the burdens.

Contrary to what the opponents would have you believe, the Convention won't affect bakers and candlestick makers. It won't put companies out of business. And it won't keep life-saving pharmaceuticals off the market. Those arguments are not true.

The Chemical Weapons Convention protects vital commercial interests. I know because we helped write the provisions related to confidential business information. I know because we helped design the reporting forms. And I know because we helped develop inspection procedures that protect trade secrets while providing full assurance that chemical weapons are not being produced.

I've heard some suggest that the Chemical Weapons Convention is going to stifle research into cancer-fighting drugs, or prevent their manufacture. But if you ask the people who make those products, they will tell you the Convention will not have an adverse effect on their operations.

The treaty doesn't prohibit legitimate medical, pharmaceutical, agricultural and industrial uses of the listed chemicals. It provides a built-in process for easing restrictions on the most promising new uses of chemicals affected by the treaty. Remember, our industry helped negotiate this treaty. We accomplished our goal.

Some treaty opponents have said we're better off if the United States never becomes a full party to the CWC. "We'll save millions," they say; "There'll be no impact if the U.S. is not a player." And as before, this argument is not true.

Ironically, our industry will suffer the greatest impact if the United States *does not* ratify the Convention. Chemical manufacturers are America's single largest exporting exporter. We exported over $60 billion in products and technology last year, with a $20 billion trade surplus. It is no exaggeration to say that our industry is the *world's* preferred chemical supplier.

If the United States does not ratify the treaty, that status will change. Our largest trading partners are also party to the Convention, and will be forced to apply trade restrictions to chemicals that originate here, or that are being shipped here.

Even if the restrictions only apply to a small portion of the overall chemicals trade, our customers have an incentive to make life as easy as possible. They'll shop where there are no limits. The result? Potentially hundreds of millions of dollars of lost sales, for no other reason than the United States is not part of the CWC.

In short, the Chemical Weapons Convention makes good business sense and good public policy. It is the next best step to prevent terrorists from getting their hands on materials to make chemical weapons. The treaty gives 160 governments—including the United States—the tools to uncover and act against the illegal chemical weapons efforts of rogue countries. Every American, and every American soldier and sailor, will be safer when the Convention is in force. Attached to this statement is a summary of the very good reasons why the Convention should be approved.

Our request to this Committee is simple. Pass the Chemical Weapons Convention. Be assured of the chemical industry's support for this important agreement.

CHAPTER SIX
The Future of Chemistry

W hat will chemistry and chemical technology look like in the next decade and beyond? To some extent, that question is easy to answer. Chemists and chemical engineers will almost certainly be working on problems that have challenged them for years, if not decades. For example, what are the "rules" and mechanisms by which proteins fold? Or how can an economically efficient fuel cell be manufactured? Or what are the biochemical mechanisms by which cancer develops, and how can they be stopped? Some of the work described in chapters 1 and 2 is of this type, modest steps forward in the solution of large research problems.

An important focus of new research will undoubtedly be based on questions that have arisen and discoveries that have been made only recently. There can be little doubt that the discoveries of nanotubes and dendrimers, for example, will lead to a fruitful line of research in the next decade. One can seldom know, however, which of the discoveries of seemingly lesser importance described in chapters 1 and 2 may lead to large, new areas of chemical research and development.

Finally, we can anticipate startling new breakthroughs in chemistry and chemical engineering that no one in the 1990s had imagined or predicted. Such is the nature of the chemical sciences—indeed, of all sciences—that one can never know what new feature of nature will be revealed by current or future research. Especially as chemists and chemical engineers begin to probe matter at its most basic level of atoms and molecules, the probabilities of striking new discoveries seem to grow exponentially.

So one can predict a future of chemistry that includes slow growth towards the solutions of existing problems, studded with a few sparkling breakthroughs that provide instant and broad insights into these problems or that suddenly and dramatically illuminate new lines of research to be pursued.

CHEMISTS PREDICT THE FUTURE

Chemists and chemical engineers are well aware of the streamlike nature of their discipline. They may not be philosophers of science, analyzing the long history of chemical thought and the forces that drive research, but they clearly understand that solving a problem in the chemical sciences means, to a large extent, discovering new questions to ask. They understand that chemical research does not lead to a final "answer," but is, instead, a never-ending sequence of question-answer-question-answer-question. . . . For this reason, virtually every chemical paper begins with a review of previous research (a summary of previous answers) and ends with a series of recommendations for new research (a summary of new questions).

On the other hand, chemists and chemical engineers seldom engage in long-term, global prognostications about the futures of their disciplines. The challenge is simply too daunting, as there are too many uncertainties and potential twists and turns in chemical research.

Exceptions to this method of inquiry do exist, however. Special events in human history or in the chemical sciences may cause chemists and chemical engineers to pause, attempt a broad assessment of developments in their science, and make an effort to imagine future directions in chemistry. Currently, the special event is the impending end of the second millennium—the approach of the year 2000. As the 1990s draw to a close, chemists and chemical engineers, like many other groups, have been taking time out of their daily routines to talk with each other about what the new decade and the new century might hold for their disciplines.

One of the most important products of this process of introspection was a report entitled *Technology Vision 2020*, which was issued jointly in December 1996 by the American Chemical Society, the American Institute of Chemical Engineers, the Chemical Manufacturers Association, the Council for Chemical Research, and the Synthetic Organic Chemicals Manufacturers Association. It summarized more than two years of work by a study group, the Technology and Manufacturing Competitiveness Group, that "began a study of the factors affecting the competitiveness of the [chemical] industry in a rapidly changing business environment" and set out to develop a vision for its future (*Technology Vision 2020*, p. 7)

By happenstance, the approach of the twenty-first century also marked the 75th anniversary of one of the profession's oldest and most esteemed journals, *Chemical & Engineering News*. In celebration of this event, the journal devoted its entire 12 January 1998 issue to a retrospective of its preceding 75 years and some projections as to the possible future of the chemical sciences over the next 25 years. The projections were summarized in a long article entitled "Chemistry's Golden Age," a report to which a dozen of the nation's leading chemists contributed. The rest of this chapter is devoted to a summary of these two visions of the chemical sciences at the dawn of the new century.

Technology Vision 2020

Technology Vision 2020 (hereafter, TV2020) focused on four general areas involved in the development and use of chemical products: new chemical science and engineering technologies, supply chain management, information systems, and manufacturing and operations. The last three general areas are briefly summarized below, followed by a somewhat more detailed discussion of the first area: new chemical science and engineering technologies.

Supply chain management, the second general area, refers to all the steps involved in obtaining materials from a supplier; handling, transporting, and storing those materials; and overseeing inventory. TV2020 suggested a number of changes by which this series of steps could be made more efficient.

Information systems constituted the third major topic covered by TV2020. The term *information systems* refers to all of those methods by which chemical data are "turned into information and used, managed, transmitted, and stored" (TV 2020, p. 16). The report recommended the development of more rational integrated systems for dealing with data; greater cooperation between government and the chemical industry, and within the chemical industry itself, in order to yield improved systems for molecular modeling; and the simulation and development of decision-making tools capable of being used with both multinational and multiproduct enterprises.

TV2020 recommended five changes in the area of manufacturing and operations: (1) improving the ability of chemical industries to recognize and respond to customer needs; (2) enhancing production methods to ensure process and product safety and to reduce deleterious environmental effects; (3) increasing the efficiency of plant production and operation; (4) improving elements in the supply chain process; and (5) expanding the industry's capability for global operations and commerce.

For readers of this book, the area of greatest interest in TV2020 is probably its vision of directions for chemical science and technology. The report identified three major challenges for the chemical sciences in coming decades: chemical synthesis and catalysis; bioprocess and biotechnology; and materials technology. In addition, the report recommended changes in certain "enabling technologies," support fields that are essentially distinct from, but whose contribution is essential to, the development of chemical science and technology.

One of the enabling technologies is *process science and engineering technology,* a term that refers to all those processes that make possible the conversion of discoveries in chemical research to practical applications for consumer products—processes such as chemical engineering, plant design, scale-up of operations, and plant construction and operation.

Another enabling technology is chemical measurement, a field that becomes increasingly important as chemists and chemical engineers develop the capability of working with materials at increasingly finer scales. A third enabling technology includes the computational technologies now used, according to the report, "in almost every aspect of chemical research, development, design, and manufacture" (TV 2020, p. 15). The most important applications of computational technologies to the chemical sciences, the report goes on, are in the fields of computational molecular science, computational fluid dynamics, process modeling, simulation, operations of optimization, and control.

Chemical Synthesis

To a chemical engineer, *chemical synthesis* is the process by which certain raw materials, such as petroleum, natural gas, minerals, and biomass, are converted to useful products, such as drugs and medicines, dyes, synthetic fibers, plastics, and synthetic building materials. One of the best-known examples of chemical synthesis is probably the conversion of ethylene (ethene) gas obtained from petroleum to the polymer polyethylene (polyethene). Polyethylene itself can then be transformed into a variety of forms with different molecular structures, different properties, and, hence, different applications.

The key element in many synthetic processes is the use of a catalyst, a chemical that speeds up or slows down the rate of a chemical reaction. The polymerization of ethylene, for example, occurs far too slowly on its own to have any commercial value. A judicious choice of catalyst and reaction temperature, however, greatly increases the rate of polymerization, which essentially makes the production of polyethylene a commercial reality.

TV2020 suggests improving synthetic and catalytic operations by

- finding new techniques of synthesis that cross disciplinary lines and that draw from knowledge and advances in fields such as biology, physics, and computational methods;
- developing new catalysts and reaction systems that can be used to make products from a greater range of readily available raw materials and can be used to recycle the products of such systems, thus improving the environmental and economic qualities of such products;
- improving methods for the synthesis of both multifunctional and highly specific molecules and products that involve processes such as self-assembly, biologically based pathways, and greater use of mixed organic-inorganic compounds;
- expanding research on nontraditional reaction media, such as gas-phase systems, supercritical fluid, less well studied aqueous environments, and reactions carried out in bulk.

Bioprocess and Biotechnology

Bioprocess and biotechnology already account for a very significant portion of the chemical technology market, producing an estimated 30 billion pounds of products per year and nearly $10 billion in annual sales. As TV2020 pointed out, however, much of the potential in this field remains untapped. For example, the report claimed that "99 percent of the microbial world has been neither studied nor harvested to date" (p. 33), suggesting an enormous promise for new biocatalysts. The improvement of current bioprocess and biotechnology, TV2020 argued, depends on identifying these new biocatalysts, isolating the enzymes they contain, clarifying the mechanisms by which previously studied enzymes work, and developing sequential enzymatic pathways by which products can be made more efficiently.

Materials Technology

Developments in materials technology are, to some extent, well under way. Chemists have already learned how to replace natural materials, such as wood, metal, stone, glass, and natural fibers, with synthetic products that often have properties superior to those of the natural materials. Some examples of these synthetic products include specialized polymers, "smart" materials, thin-film surface products, and self-assembling materials.

The authors of TV2020 place the highest priority in materials technology on finding ways to predict the properties in advance of new materials "from the molecular level through the macroscopic level" (p. 36). As this ability is enhanced, scientists and engineers will greatly increase their abilities to efficiently design and create new products, from both novel and natural materials, for very specific applications. The authors also identify specific kinds of materials on which work should be advanced: sensors for chemical process industry use; materials with improved environmental stability and durability; electrical and optical materials; advanced "smart" materials; materials compatible with living systems; materials for use in high-temperature environments; and specialized membranes for use in separation processes used in chemical processing, medical applications, and other fields.

Reference:

Technology Vision 2020: The U.S. Chemical Industry. Washington, DC: American Chemical Society, American Institute of Chemical Engineers, Chemical Manufacturers Association, Council for Chemical Research, and Synthetic Organic Chemical Manufacturers Association, December 1996.

"Chemistry's Golden Age"

The *Chemical & Engineering News* panel provided a view of chemistry in the year 2023—the 100th anniversary of the journal—that can be characterized as stunningly optimistic and exhilarating, with a sound basis in existing research trends and societal needs. One focus of the panel's discussion was the future of certain structural features of the chemical sciences, such as employment, publishing, and funding.

Members of the panel included John D. Baldeschwieler, California Institute of Technology; Allen J. Bard, University of Texas, Austin; Jacqueline K. Barton, California Institute of Technology; Ronald Breslow, Columbia University; Theodore L. Brown, University of Illinois, Urbana-Champaign (emeritus); Barbara Imperiali, California Institute of Technology; Robert S. Langer, Massachusetts Institute of Technology; Koji Nakanishi, Columbia University; Daniel G. Nocera, Massachusetts Institute of Technology; Douglas Raber, Board on Chemical Science & Technology, National Research Council; Stuart A. Rice, University of Chicago; and Richard E. Smalley, Rice University. Richard Zare, National Science Board and Stanford University, and Stephen J. Lippard, Massachusetts Institute of Technology, also "contributed their thoughts about the future of chemistry" (p. 143).

For example, panel members debated as to whether departments of chemistry would even continue to exist in colleges and universities 25 years into the future. One view was that chemistry had and would become so thoroughly incorporated into the other sciences that chemistry departments would no longer be needed. Instead, chemists would find themselves in departments of molecular biology ("chemical biology") or materials science ("materials chemistry"). Other panelists disagreed. Chemistry is just too fundamental a subject for it to be entirely discarded as a distinct discipline, they said. Panelist Daniel Nocera opined that "chemists make new things, and we study reactions. That's the core of this profession. And that can never go away" (p. 144).

As in the TV2020 report, the journal's panel expected the greatest advances in chemistry over the next generation to occur at the interfaces between chemistry and biology and chemistry and materials science. In one of the most startling and pregnant predictions, Richard Zare argued that one of chemistry's greatest challenges in the twenty-first century would be to create life—a goal he thinks is attainable. As other panel members pointed out, progress in achieving that goal has already been made, and they expected that chemists would be able to manufacture working cell membranes by 2023 and, soon thereafter, to build functional cells containing such membranes.

Important medical breakthroughs were also expected as the result of advances in the chemical sciences. Jacqueline K. Barton anticipated the discovery and invention of small molecules, such as those now used in gene therapy and protein therapy, to treat diseases. "We can use small molecules as the basis of a whole set of general tools," she said, "general solutions that we can use to fight different diseases" (p. 145).

Chemistry will also be used to provide a better understanding of the nature of neurons and neuronic systems, thus improving our knowledge of processes such as memory and intelligence. Zare predicted that one result of this research might be an improved understanding of the nature of sleep, leading to a form of medication that might make it possible for humans to function efficiently with only one hour of sleep a day.

In the area of materials science, panelists were especially excited about the possibilities of improving our understanding of chemical synthesis and catalysis to a point where physical experimentation might no longer be necessary. Computational techniques might be advanced to a point where specific reactions could be traced step-by-step, and final products could be predicted without ever manipulating materials in a laboratory.

One consequence of such a development might be a vastly improved method of choosing, identifying, and using catalysts. Instead of relying on the trial-and-error method now used in catalysis research, chemists could select previously identified candidates stored in computer libraries for use in specific reactions.

The need for improved analytical techniques was also raised by panel members. Recent research has pointed out the need for more refined and specific methods of analysis, including those that can be used at the level of single molecules. Richard Smalley described his own vision of "pure spectroscopics for individual molecules" (p. 149) that would make possible, for example, NMR (nuclear magnetic resonance) analysis of a single molecule.

Important advances in materials science, according to some panel members, could be expected at the interface of organic chemistry and electronics. The development of nanotube technology, for example, has opened the possibility of creating virtually any kind of electrical system—on either a micro- or macroscale—by using organic materials. Smalley predicted that the year 2023 would find organic chemists making electronic devices rather than pharmaceuticals or polymers.

Over the next 25 years, chemistry will also make significant contributions to the everyday lives of people. For example, chemists have already learned how to make a nutritious, synthetic fat. They can also be expected to develop other materials that will help in what may be the world's number one problem: food supply.

Chemists will also make important contributions to solving the world's energy problems. Allen J. Bard predicted that "twenty-five years from now the internal combustion engine will be found in museums" (p. 150). Transportation systems will be powered instead, he said, by vastly improved battery technologies and efficient fuel cells. These changes will be necessitated, panel members felt, not only by the dwindling supply of petroleum products (which are, in any case, more valuable as raw materials than as fuels), but also by growing problems of climate change created by the combustion of fossil fuels.

The panelists' overall outlook regarding the status of chemistry and chemical engineering in the next 25 years was summed up by moderator Rudy Baum who described the panelists as "irrepressible in their enthusiasm for what chemistry can accomplish" (p. 151). Stephen J. Lippard expressed the panel's enthusiasm and optimism in the following statement: "We don't just study the natural world, we make new molecules, new catalysts, new compounds of uncommon reactivity. Part of our subject allows us to be creatively artistic through the synthesis of beautiful and symmetric molecules. Our ability to rearrange atoms in new ways allows us a tremendous opportunity for creation that other sciences do not have" (p. 151).

Reference:
Baum, Rudy M., "Chemistry's Golden Age," *Chemical & Engineering News*, 12 January 1998, 143–62.

CHAPTER SEVEN
Career Information

At one time, it was relatively simple to describe what a "career in chemistry" involved. Lavoisier, Berzelius, Gay-Lussac, Dalton, Priestley, Scheele, Cavendish, and the other founders of modern chemistry were relatively well-to-do men who could afford to build, operate, and supply their own laboratories. These chemical giants often pursued a wide variety of chemical topics in the privacy of their own homes.

Over a period of two centuries, careers in chemistry have grown and become far more complex. Today, chemists work in a wide variety of academic, industrial, governmental, private, and other institutions. Their studies usually are focused on a very specific aspect of chemistry, such as the structure of a particular molecule, the development of techniques for analysis, or the detailed examination of a single class of chemical reactions.

To fully describe the range of career opportunities available to chemists and chemical engineers today would require a chapter much larger in scope than is possible here. Instead, the purpose of this chapter is to provide an overview of the opportunities available to individuals with an interest in chemistry and to offer a closer look at some specific job opportunities.

In general, work opportunities in chemistry can be classified in two ways: by the type of organization by whom one is employed and by the field of chemistry in which one is employed.

According to studies conducted by the American Chemical Society (ACS), of all chemists employed in the late 1990s, about 60% work in industry; 24% work in schools, colleges, and universities; 9% work in government; and 7% work in other settings.

CHEMICAL CAREERS IN INDUSTRY

The kinds of jobs held by chemists in industrial settings can be described in general as basic or applied research and product development, and sales, marketing, and technical service.

Basic or Applied Research and Product Development

Research work can be classified as *basic* or *applied*. Basic research is research conducted to discover fundamental information about the nature of matter. For example, a chemist might be interested in finding out how the physical, chemical, and biological properties of a compound change when the molecules of that compound are slightly altered. The research is not initiated because the chemist wants to develop some new product that will be valuable to humans, but simply because he or she is interested in learning more about a particular compound. Of course, this research *might* lead to the discovery of a new commercial product. In fact, basic research often leads to such practical applications. But the practical applications themselves are not the original purpose of the research.

Applied research, by contrast, is aimed at solving specific chemical problems. A new compound discovered as the result of basic research can be put to practical use if and only if some chemists solves a whole variety of practical problems. The way in which a particular compound behaves in a test tube or a distilling flask may be very different from the way it behaves in the real world. The goal of the development chemist is to translate the discoveries made by basic research into useful products for the real world.

Yet a third field of industrial chemistry is product development. Compounds and procedures that have been discovered and developed must eventually be produced by some large-scale process. A production chemist is involved in the design and operation of plant equipment used to manufacture compounds, ensures quality control, and may be responsible for environmental regulations established by various levels of government.

Sales, Marketing, and Technical Service

Many people who start out as chemists find that they are not interested in laboratory work or teaching. They discover that they are more interested in business, human resources, marketing, or other such fields that often overlap with chemistry.

For example, each year hundreds of new chemicals are produced for the marketplace. These include cosmetics, paints, drugs and medicines, fabrics, plastics, and other synthetic materials. Getting these products to the marketplace and selling them to consumers requires the efforts of many kinds of workers, most of whom need to know at least something about chemistry.

For example, companies employ specialists in marketing who know something about the potential demand for certain kinds of chemicals, future trends in that demand, and ways in which new chemicals can be marketed. While such specialists may not *have* to know very much chemistry, a chemical background can be very useful in plotting such information and laying out strategies.

Anyone involved in the sales of new chemicals, on the other hand, will certainly have to be familiar with the new chemical products for which he or she is responsible and to be able to talk with prospective customers about the chemical, physical, and biological properties of those products. As an example, salespeople who work for drug companies must be very knowledgeable about their products when they talk with physicians, nurses, medical practitioners, and other medical specialists. Booksellers of chemistry texts need to know a fair amount of the subject of the books if they are to talk intelligently to prospective buyers.

A growing field in industry is technical services. Many companies now buy and use very expensive, complicated chemical instruments. Companies that sell such instruments need to employ knowledgeable people who can help design such equipment, write instruction manuals for it, teach company employees how to use it, and "trouble-shoot" when problems develop with the equipment in the field.

CHEMICAL CAREERS IN ACADEMIA

Chemists usually choose to work in schools, colleges, and universities for one of two reasons: the opportunity for pursuing their own research interests or the satisfaction of teaching.

Educational institutions, especially the larger ones, can provide chemists with the time and setting needed to carry out their own chemical research. Professors at major universities commonly devote the majority of their time to research, and spend only a relatively modest amount of time on teaching, advising students, serving on committees, and doing other academic chores. Professors at smaller universities and most colleges, however, may be more likely to spend most of their time teaching. They may still find some time to carry out their own research, but that research is seldom the major focus of their responsibilities at the college or university.

Much of the teaching in chemistry done at large universities is often the responsibility of teaching fellows. Teaching fellows are graduate students who are working on their own advanced degrees. They earn a living during this period of time by teaching large classes, running small discussion sections, serving as laboratory instructors, or carrying out other teaching responsibilities.

Many colleges and universities also hire technical specialists who assist with the "hardware" aspect of chemistry at the institution. They may operate the chemical stockroom and take responsibility for maintaining an inventory of equipment. Or they may be in charge of the operation and maintenance of complex instruments and equipment that are now part of most academic chemical laboratories.

Many people with an interest in chemistry are more attracted to the teaching aspect of the subject than to research, development, or some other aspect of chemistry. These people may choose to become high school chemistry teachers, middle school science teachers, elementary school science supervisors, or school curriculum directors—fields in which they can combine their interests in chemistry and education.

CHEMICAL CAREERS IN GOVERNMENT

The federal and state governments provide employment opportunities in virtually every conceivable field of chemistry. The U.S. Environmental Protection Agency and the U.S. Department of Agriculture, along with their state counterparts, employ chemists to conduct tests for the presence of certain chemicals in water supplies, soil, plants and animals, and foods we eat. The Department of Energy employs chemists who are interested in the study of fuel composition, combustion, alternative energy sources, and related subjects. The National Oceanic and Atmospheric Agency employs chemists who study the chemical composition of the oceans and the atmosphere and who study changes that occur in these environments as a result of human activities. The U.S. Food and Drug

Administration employs chemists who analyze foods to determine their nutritional value and who study the effects of various food additives on the health of humans.

SPECIAL CAREERS IN CHEMISTRY

Many people who want a career in chemistry are motivated by the opportunity to work in a laboratory, discovering new materials or finding new uses for familiar materials. Others who enjoy working in chemistry are more interested in putting their love of chemistry to work in nonlaboratory, nonteaching settings. As an example, some institutions employ librarians with a special expertise in chemistry. A large chemical company, such as Dow Chemical, for example, may hire a number of librarians who can answer questions and find resources for the thousands of chemists who work at the company.

Some chemists also choose careers as financial analysts in which they use their knowledge of chemistry to predict future patterns of growth and earnings in chemical companies. Such individuals have a double responsibility since they must have expertise not only in the field of investment finances, but also in the field of chemistry itself.

Chemical patent attorneys are lawyers with a special interest and background in chemistry. Their job is to follow the development of new substances to see if they are patentable and, if so, whether such patents already exist. In some cases, chemical patent attorneys have the opportunity to work at the forefront of the legal profession. The discovery of new lifelike substances, for example, raises questions as to what is "natural" or "synthetic" and what can or cannot be patented by its discoverer.

There is hardly a field of human endeavor in which chemists cannot have input. For example, the art world is often interested in the authentication of a painting. Is a painting signed by Leonardo DaVinci really a work by that great artist, or is it an excellent forgery? Chemical analysis of the paints used in the work and of the canvas on which it is painted can help answer such questions.

Specialized demands such as this one have created a market for chemical consultants. Chemical consultants are people who prefer not to work in a regular 8-to-5 job, but to make themselves available for working on special chemical problems. In some cases, consultants specialize in some specific field, such as the analysis of works of art. In other cases, they tend to be generalists who take on a wide variety of chemical problems encountered by private industry or governmental agencies.

FIELDS OF CHEMISTRY

Traditionally, chemistry has been divided into four large fields: organic, inorganic, analytical, and physical chemistry. This method of categorizing chemists and the work they do is now much oversimplified. However, a great many chemists still describe themselves in one of these four general terms.

Organic chemists study the structure and behavior of compounds that contain carbon. The field of organic chemistry is an enormous one, with more than 90% of all known chemicals being organic. A vast array of commercial products, from drugs and medicines to pesticides and synthetic fertilizers to synthetic fibers and dyes to synthetic rubber, are made from organic compounds.

An important subdivision of organic chemistry is biochemistry, the study of chemical compounds found in living organisms. The field of biochemistry is one of the many fields that has grown up at the intersection of two or more fundamental sciences. Other examples include geochemistry (the study of chemical compounds found in the Earth), astrochemistry (the study of the chemical compounds found in stars and interstellar matter), and physical chemistry and chemical physics (the study of the physical properties of atoms, molecules, and other chemical species).

Inorganic chemists study compounds of all elements *except* carbon. These elements include such well known substances as oxygen, hydrogen, iron, tin, sulfur, aluminum, and magnesium, as well as less well known elements such as lanthanum, cerium, xenon, dysprosium, tantalum, thulium, and astatine.

Analytical chemists determine the structure and composition of compounds and mixtures as well as develop and operate instruments and techniques for carrying out such analyses. For example, an analytical chemist might be engaged in finding out the percentages of various elements present in a newly discovered mineral, in determining the amount of pesticide present in runoff from an agricultural plot, or in analyzing the composition of alloys formed in the reaction between iron and other elements.

Physical chemists use the principles of physics to understand chemical phenomena. They frequently use both mathematical and mental models of phenomena that are too small to study firsthand. Some areas in which physical chemists are interested are the structure and properties of atoms and molecules, quantum mechanical models of electron behavior, thermodynamics of chemical systems, surface chemistry and catalysis, and electrochemistry.

Chemical Engineering

The discipline of chemical engineering involves applying the principles of the chemical sciences to everyday problems in the production and use of materials. Chemical engineers are involved at virtually every step, from the discovery of new materials and new chemical systems to the production and sale of those materials and systems.

The American Institute of Chemical Engineers lists six primary areas in which chemical engineers are employed.

- *Research*: Some chemical engineers work directly with research chemists to develop new products and new techniques. For example, they might assist chemists in the development of a new kind of plastic or a new way to produce an existing plastic.
- *Design*: Chemical engineers are responsible for developing the methods and materials needed to translate a new discovery in the chemical laboratory to a commercially successful product.
- *Development*: The field of development involves finding ways to improve new or existing processes. By developing new and more efficient ways to produce synthetic fertilizers, for example, chemical engineers made a major contribution to the improvement of agricultural processes throughout the world.
- *Production*: The physical processes by which chemicals are manufactured involve an almost endless list of technical questions, such as how much of each raw material to use, how to ensure that all chemical reactions occur as efficiently as possible, how to move chemicals from one step of an operation to the next, and how to ensure that a manufacturing process proceeds as safely as possible.
- *Technical sales:* Some chemical engineers choose to apply their expertise in chemistry to solving practical questions involving the use of chemical products. For example, some products may not perform in the real world the way they did in the laboratory or the way a company expected that they would. A chemical engineer employed in technical sales would be involved in solving problems of this type.
- *Management*: Chemical engineers who particularly enjoy working with people may choose to work in some form of management. In this position, they may be responsible for making business and policy decisions, training new employees, and solving human problems rather than those involving chemical reactions.

Chemical engineers find work in a wide variety of situations, such as the food, electronic, textile, pulp and paper, rubber, metal, cement, plastic, pharmaceutical, and aerospace industries. Government agencies such as the Department of Energy and the Environmental Protection Agency also have positions for chemical engineers.

According to the *Occupational Outlook Handbook*, about 7 out of every 10 chemical engineers are employed in chemical industries, petroleum refining, and related industries. The remaining 3 out of 10 chemical engineers work for government agencies, for engineering services, in research and testing organizations, or as consultants. About 50,000 jobs in chemical engineering are currently available.

As with chemists, chemical engineers often hold highly specialized jobs, such as working on the chemical aspects of electronics or drug production. In many cases, chemical engineers are required to be familiar with a greater range of scientific fields than are chemists. For example, they may need to be proficient in physics, mathematics, and other fields of engineering.

Chemical Technology

Chemical technicians are men and women who conduct some of the more routine aspects of chemical research and development. Such individuals have probably been around since the dawn of the chemical sciences. During the alchemical era, for example, the master alchemist needed someone to keep the fire going, grind chemicals, clean up the laboratory, and perform any number of similar activities.

Chemical technicians today perform tasks similar, though far more sophisticated and complex, to those conducted by alchemical assistants. They set up and operate laboratory equipment, order and keep inventories of chemical supplies, repair and maintain equipment, monitor experiments and record results, and are responsible for dozens of simpler tasks assigned to them by senior chemists and chemical engineers.

As an example, a chemical technician working in the field of environmental chemistry might be expected to store testing equipment, prepare equipment for field tests, conduct tests of air or water quality, record the results of those tests, repair damaged equipment, and report on the results of experiments to senior chemists.

Perhaps the biggest change in the job descriptions of typical chemical technicians is the increased use of sophisticated instruments in their work. Not so many years ago, most of the work done by chemical technicians was done manually: samples were individually weighed, and mixtures were separated and analyzed by hand. Today, much of the routine work in a laboratory is carried out by some kind of instrument.

The use of such instruments greatly simplifies the work of the technician but demands a much higher degree of understanding and mechanical skill than has traditionally been the case. In the future, those employed in this field will probably be expected to have an even better understanding of and greater experience in the use of instruments.

In many cases, the tasks performed by a chemical technician are limited, to a large extent, only by his or her ambition. As one becomes skilled at the use of equipment, he or she may be assigned more complex tasks and, perhaps, even be given the chance to do some original research. It is not unusual for a chemical technician to begin his or her career working under the close supervision of one or more senior technicians or senior chemists and chemical engineers. Over time, however, the technician will be given more responsibility to work on his or her own and, ultimately, may be assigned a supervisory position.

The Bureau of Labor Statistics (BLS) estimates that about 231,000 science technicians were employed in 1994. About a third of that number worked in manufacturing, primarily in the field of chemistry. The federal government itself employed about 17,500 science technicians, primarily in the Departments of Defense, Agriculture, and Interior.

BLS predicts that job opportunities for chemical technicians will increase at about the same rate as the average for all occupations through the year 2005. Those employed in the field can probably be expected to need a better understanding of and greater experience in the use of instruments.

EDUCATION AND TRAINING REQUIREMENTS

The vast majority of jobs in chemistry and chemical engineering require at least a bachelor's degree in chemistry. A bachelor's degree alone may qualify a person for a position as a high school chemistry teacher, a laboratory technician, a sales representative, or another similar-level position. For anyone interested in research or positions in management and supervision, however, an advanced degree—usually a Ph.D.—is required.

Thus, the first step for someone who has chosen a career in chemistry is to complete a bachelor's degree in chemistry. The next step is to gain experience by working in an entry-level job or to go on to a program leading to a master's degree. The master's degree greatly increases the range of jobs open to a person. Completion of the master's degree again leads to a choice of work experience or graduate work leading to a Ph.D.

Most chemists interested in a career of research continue their education beyond the Ph.D. and complete one or more years of postdoctoral studies. During this phase of their education, they often become members of research teams to further develop their skills in research. Eventually, postdoctoral studies are followed by a position in industry, academia, or government, as outlined above.

Many chemists are happy to stay at one particular level throughout their lives. They may be happy to do routine laboratory work that requires no more than a bachelor's degree. Or they may be satisfied to continue doing basic research in the relative privacy of their own laboratories. Many other chemists, however, become interested in jobs outside of the field of research, and strive to become administrators at large universities or executives of companies involved in chemical work.

For individuals who are not interested in pursuing a full, four-year college course in chemistry, a career as a chemical technician may be an option. Most chemical technicians are expected to have at least two years of training in the field. That training is offered by community and junior colleges, some four-year colleges, and special training schools. For anyone who has hopes of later continuing his or her education in chemistry, the community, junior, or four-year college option may be a better choice than a special training or technical school. Courses taken in such programs can often be transferred for credit to four-year chemistry programs, while courses taken at technical schools often cannot be transferred.

EMPLOYMENT OPPORTUNITIES

The BLS has predicted that job opportunities for chemists will grow about as fast as the average for all occupations in the United States through the year 2005. The BLS predicts that job opportunities for chemical engineers, on the other hand, will grow very little during that same period of time. These predictions have somewhat limited value, however, since opportunities are likely to vary considerably within both fields. For example, chemists specializing in research on chemicals used in the field of electronics are likely to be more in demand than those working in some other field. For chemical engineers, BLS has predicted that job opportunities are likely to grow fastest in the fields of specialty chemicals, pharmaceuticals, and plastics materials.

Where Are the Jobs?

One of the best ways of finding out about careers in chemistry is by reading the want ads published in daily newspapers and professional journals. The following job descriptions have been taken from such publications in the past year:

- Colgate-Palmolive Company was looking for a chemist with a Ph.D. and 0–3 years of experience to work on the physical interactions between active materials, such as fragrances and antibacterials, with product components such as polymers and surfactants.
- Geneva Pharmaceuticals, Inc., advertised for a Director of Analytical Research and Development with a Ph.D. in analytical chemistry, physical organic chemistry, pharmaceuticals, or the equivalent. The person sought would be expected to "direct the development and validation of analytical methods and specifications for all new products, pharmaceutical raw materials, and dosage forms" and to direct analytical efforts designed to meet the demands of various regulatory bodies.
- Human Resources Groundwater Analytical, Inc., was seeking an environmental chemist with at least a B.S. in chemistry, knowledge of methods and analytical instrumentation demanded by the U.S. Environmental Protection Agency, five or more years of experience in the field, and various interpersonal skills.
- Chemglass, Inc., advertised for a sales representative with a background in chemistry for developing new sales accounts and maintaining existing accounts with research chemists in a variety of occupational positions.
- Creo Products, Inc., was looking for a Chemistry Senior Researcher who would be responsible for putting together a team of chemists and chemical engineers whose task it would be to develop a new kind of photoresist that can be applied directly to metallic surfaces. Two positions were available for two parallel groups, one in Canada and the other in Israel.
- The Southern Regional Research Center of the U.S. Department of Agriculture's Hawaii Agricultural Research Center was searching for a Synthesis Associate, with a M.S. or Ph.D. The candidate's job would be to prepare monomers and polymers of sucrose.
- The Swiss Federal Institute of Technology in Zurich was seeking a person to fill the position of Professor of Biochemical Engineering. The successful candidate would be expected to apply basic engineering principles to biological systems in a variety of courses in chemistry and chemical engineering. The candidate would also be expected to be able to "cooperate with both academic and industrial colleagues."

- The New York City Police Department was seeking candidates for the position of Criminalist, Level II, III, or IV. Qualifications for the three levels involved various combinations of a B.S. or M.S. in criminalistics, forensic science, chemistry, biology, biochemistry, physics, or related fields and years of experience.
- The National Science Foundation was looking for a Program Director (chemist) to be responsible for the review of proposals and the management of portfolios at the interface of chemistry and materials research. The position required a Ph.D. or equivalent experience in chemistry or a related field.
- Nature's Sunshine Products, Inc., was seeking a Research Investigator who would be responsible for the development of techniques for the analysis of natural products. The position required a Ph.D. in analytical chemistry or equivalent and involved the supervision of three other Ph.D./M.S. scientists.
- Bend Research, Inc., in Bend, Oregon, was looking for chemists or chemical engineers with M.S. or Ph.D. degrees to work on new kinds of materials and processes for delivering drugs to the human body.
- The Yale University School of Medicine was advertising for an operator of NMR (nuclear magnetic resonance) equipment used in the study of glucose metabolism in diabetes. Candidates were expected to have at least a B.S. and experience in using NMR equipment.
- The City College of New York was searching for a new faculty member with experience in the use of computers to model biochemical reactions, to teach at both undergraduate and graduate levels, and to develop proposals for external funding of research programs.
- North Carolina State University was looking for someone with a degree in lignin biochemistry or chemistry to work on the study of quantitative variations in the lignin composition of woody plants and grasses. The person hired would be expected to work with molecular biologists, quantitative geneticists, and lignin chemists.
- The University of Botswana was seeking candidates for the positions of Lecturer in Organic Chemistry and Lecturer in Physical Chemistry. Candidates were expected to have at least a master's degree, but preferably a Ph.D. They would be expected to teach both graduate and undergraduate courses and to develop an active program of research in their own fields of special interest.

Compensation Patterns

The American Chemical Society (ACS) conducts annual surveys of salary patterns for chemists. These results are published in the society's publication *Chemical & Engineering News*. In its 1997 report, ACS reported that the median salary for all chemists in the United States was $63,000. That median differed considerably according to the field of employment. For industry, the average was $67,000; for government, $61,800; and for academia, $54,000. The median also differed depending on the highest degree held by an employee. For those with a bachelor's degree, the median pay in 1996 was $49,400; for those with a master's degree, $56,200; and for those with a Ph.D., $71,000. As in most fields, median pay also differed substantially for men and women, with the former averaging $66,600 per year and the latter averaging $50,200 per year. (Also see data on this topic in chapter 8.)

REFERENCES AND RESOURCES

A number of useful resources provide additional and more detailed information on careers in chemistry. Information on several of such resources follows.

Books, Pamphlets, and Brochures

Blumenkopf, Todd A., Virginia Stern, Anne Barrett Swanson, H. David Wohlers, eds., with Michael Woods. *Working Chemists with Disabilities: Expanding Opportunities in Science*. Washington, DC: American Chemical Society Committee on Chemists with Disabilities, 1996.

> The purpose of this book is "not to turn all people with disabilities into chemists, but to open the door for those who are 'able-minded' and have a passion for science to fulfill that passion" (p. 3). The book tells the stories of 17 men and women with a variety of disabilities who are now working in industry, government, colleges, universities, and high schools. The book tells, in an engaging manner, the way in which these individuals have made successful adaptations to a wide range of careers in research, development, administration, supervision, and teaching.

Cracking It! Stockport, England: Training Publications, Ltd., 1997.

> This book suggests ways in which women can break through the "glass ceiling" in a variety of professional occupations, including science, engineering, and technology. Available from the publishers at P.O. Box 75, Stockport, SK4 1PH, or by fax at +44 (0) 161-474-7502.

Occupational Outlook Handbook. Washington, DC: Government Printing Office, published annually.

The *Occupational Outlook Handbook* is the premier source of information about careers in the United States. Some of the many chemistry-related careers it discusses include chemists; chemical engineers; engineering, science, and data processing managers; industrial engineers; photographic process workers; engineering technicians; science technicians; pharmacists; metallurgical, ceramic, and materials engineers; respiratory therapists; surgical technologists; dental hygienists; water and wastewater treatment; and medical record technicians. Each entry in the *Occupational Outlook Handbook* includes sections on the nature of the work; working conditions; employment; training, other qualifications, and advancement; job outlook; earnings; related occupations; and sources of additional information. The home page for this resource is at <http://stats.bls.gov/oco>.

Owens, Fred, Roger Uhler, and Corinne A. Marasco. *Careers for Chemists: A World Outside the Lab.* Washington, DC: American Chemical Society, 1997.

As the nature of work in the United States changes, chemists will have to consider a greater variety of alternatives to traditional laboratory work in research and development. This book reviews a great variety of "nontraditional" careers for those with a background in chemistry. Included are careers in business, information sciences, computer modeling and software development, art conservation, historical studies, consulting, education, finance, government, human resources, law, medicine, professional and trade associations, quality assurance and control, regulatory work, science writing, technical sales and marketing, and technical service and transfer. The book also contains a very useful chapter outlining the materials and techniques that need to be explored by anyone looking for a career in chemistry.

Woodburn, John H. *Opportunities in Chemistry Careers.* Lincolnwood, IL: VGM Career Horizons, 1997.

Woodburn has written an excellent background book on careers in chemistry with chapters that deal with the nature of chemistry, how one becomes a "rich" or "much appreciated" chemist, educational requirements, personal qualities needed in chemistry, career specialties in chemistry, securing employment, professional associations and societies, and the history and future of chemistry. VGM also publishes similar books on careers in biotechnology, science (in general), environmental sciences, pharmacy, medical technology, and energy.

Many professional chemical societies publish booklets or other materials describing career opportunities in the specific field represented by the society. As an example, the American Ceramic Society publishes a very useful pamphlet entitled "Ceramic Engineering: Career Opportunities for You." The pamphlet describes specialized fields of ceramic engineering, such as aerospace, bioceramics, automotive, information services, and ceramics for daily use. It also presents brief biographical sketches of selected ceramic engineering professionals, such as astronaut Bonnie Dunbar and IBM engineer Tony Gates. Readers interested in specific fields of chemistry and chemical engineering should refer to chapter 9 for professional associations in those fields.

Articles and Related Sources

Chemical & Engineering News is an invaluable source of information about careers. The journal periodically publishes special sections that deal with career opportunities in chemistry and chemical engineering. Those sections are particularly useful because they often provide hints as to other sources of additional or more detailed information about careers in chemistry. Examples of some useful articles are listed below.

"Alternative Careers Lure Chemists Down a Road Less Traveled," *Chemical & Engineering News*, 23 October 1995, 51–55.

>This article provides a review of the forces driving chemists to nontraditional forms of employment and includes a brief discussion of such careers.

Diana Slade, "Career Planning Resources," *Chemical & Engineering News*, 3 November 1997, 52–54.

>This section is part of an issue of *Chemical & Engineering News* devoted to the employment outlook in chemistry for 1998. The section lists books, brochures, posters, and other sources of information, as well as tips for doing a career search.

Waller, Francis J., "Technicians Make It Happen," *Chemical & Engineering News*, 15 December 1997, 53.

>In this article, the Chair of the ACS Committee on Technician Activities comments on the work of chemical technicians and explains why they have a place in the American Chemical Society.

Advertising supplements in scientific journals also often provide good overviews of the kinds of jobs currently offered in chemistry. The journal *Science*, for example, regularly carries special advertising supplements that focus on particular fields of science. Examples of such supplements are listed below.

"Diversity in the Scientific Work Force," 31 October 1997, 913 et seq.

"European Careers in Biotechnology & Pharmaceuticals Industry," 10 October 1997, 316 et seq.

"Hot Careers through 2005," 26 September 1997, 2018 et seq.

Many articles related to careers in chemistry can also be accessed via the Internet. The *Science* article listed below, for example, is one of a series linked to the journal's Internet site for young scientists, <http://www.nextwave.org>. This particular series deals with the topic "Careers in Manufacturing."

Ruediger, Nicole, "Out of College and into a Rewarding Biotech Career," *Science*, 5 December 1997, 1823.

Internet Sites

A large and growing number of Internet Web pages provide a more general or more specific discussion of the variety of careers available in chemistry. Some useful sites for obtaining information on careers in chemistry are described below.

An obvious place to begin searching for more information about careers in chemistry is the American Chemical Society Web page. ACS maintains an Office of Career Services that provides a host of aids to its members and nonmembers. The ACS Web site is at <http://www.acs.org>.

A source for information on careers in chemical engineering is the Web site of the American Institute of Chemical Engineers (AIChE) at <http://www.aiche.org>. Specific career information from AIChE can be found at <http:198.6.4.175/docs/welcome/>.

The American Association for the Advancement of Science also maintains a careers site on its Web page. For an interesting selection of "Career Perspectives," see <http://recruit.sciencemag.org/feature/job-resources.shl>.

The home page for the *Occupational Outlook Handbook* is at <http://stats.bls.gov/oco>. See p. 217 for a more complete description of the OOH.

An interesting introduction to careers that combine chemistry with fields such as biotechnology, law, cosmetics, food, forensics, textiles and fabrics, and toxicology and industrial hygiene is "The Alchemist's Lair" Web site at <http://snyoneab.oneonta.edu/~pencehe/other.html>.

Other Web sites that provide background information on careers in chemistry and that often supply links to related pages include the following:

<http://hackberry.chem.niu.edu.70/1/ChemJob>

<http://www.as.wvu.edu/chemistry/cic.html>

<http://www.canterbury.ac.nz/chem/careers.htm>

<http://www.carleton.ca/~jplogan/chem/careers.html>

<http://www.chem.duke.edu/~bonk/Careers/ChemCareers.html>

<http://www.chem.ukans.edu/chem/ugrad/careers.htm>

<http://www.chemistry.ucsc.edu/Undergrad/career.html>

<http://www.chemsite.com/cw.html>

<http://www.general.uwa.edu.au/u/chem-www/careers.htm>

<http://www.georgian.edu/chemistry/ch_jobs.htm>

<http://www.lacebark.nut.edu.au/chemistry/career.html>

<http://www.maths.unsw.edu.au/science/chemcar.html>
<http://www.st.nepean.uws.edu.au/Courses/careers/careers>

CHAPTER EIGHT
Statistics and Data

Ｔhis chapter provides data on a number of different aspects of the chemical industry, including quantities of chemicals produced, employment trends for chemists, and extraction and utilization of fossil fuels. All data apply to the United States unless otherwise indicated in a table.

RECENT TRENDS IN THE CHEMICAL INDUSTRY

Value of Shipments, 1990–1995 (in millions of dollars)						
	1990	1991	1992	1993	1994	1995[1]
Chemicals and allied products	288,184	292,326	305,371	314,907	333,249	360,026
Inorganic chemicals	23,487	23,572	24,052	23,063	21,619	23,348
Petrochemicals	118,989	115,600	117,074	119,225	131,712	146,314
Plastics, resins, and synthetic rubber	31,326	29,566	31,304	31,546	36,965	43,503
Synthetic fibers	12,884	12,581	12,862	13,293	13,366	13,495
Drugs	53,720	60,835	67,668	70,985	76,238	79,136
Soaps, cleaners, and toilet goods	41,438	41,854	43,055	46,903	46,314	47,474
Paints and allied products	14,239	14,255	14,974	160,030	17,544	18,651
Organic chemicals	65,053	63,721	63,827	63,541	68,822	75,505
Fertilizers, nitrogen	3,113	3,238	3,175	3,467	4,246	4,953

Value of Shipments, 1990–1995 (in millions of dollars) *(continued)*

	1990	1991	1992	1993	1994	1995[1]
Fertilizers, phosphatic	4,636	4,984	4,333	3,648	4,597	5,495
Agricultural chemicals	8,539	8,346	9,151	9,554	9,636	9,958
Adhesives and sealants	5,485	5,483	5,659	5,859	5,849	6,284
Petroleum refining	159,411	145,392	136,239	129,961	128,236	134,198

Source: U.S. Department of Commerce, Bureau of the Census, *Statistical Abstract of the United States, 1996*, Tables 1403–15.
[1] Estimated

Total Employment in Various Fields of Chemistry and Chemical Engineering, 1990–1995 (in thousands of jobs)

	1990	1991	1992	1993	1994	1995
Chemicals and allied products	853.0	846.0	848.0	838.0	824.0	812.0
Inorganic chemicals	92.4	95.6	94.8	85.0	78.8	n/a
Petrochemicals	254.0	252.0	256.0	255.0	252.0	n/a
Plastics, resins, and synthetic rubber	62.4	60.5	60.4	62.2	69.2	68.6
Synthetic fibers	57.8	57.4	55.4	51.6	46.9	46.0
Drugs	183.0	184.0	194.0	200.0	206.0	203.0
Soaps, cleaners, and toilet goods	126.0	123.0	123.0	124.0	118.0	116.0
Paints and allied products	53.9	51.1	51.2	50.3	50.1	49.4
Organic chemicals	123.0	125.0	122.0	121.0	112.0	111.0
Fertilizers, nitrogen	7.5	7.3	7.0	7.0	8.0	8.0
Fertilizers, phosphatic	10.5	10.3	9.5	9.4	8.5	8.4
Agricultural chemicals	17.7	16.4	16.9	16.3	15.3	15.1
Adhesives and sealants	21.4	20.9	21.1	20.9	19.2	18.9
Petroleum refining	71.9	73.9	74.8	73.1	71.7	n/a

Source: U.S. Department of Commerce, Bureau of the Census, *Statistical Abstract of the United States, 1996*, Tables 1403–15.

Number of Production Workers in Various Fields of Chemistry and Chemical Engineering, 1990–1995 (in thousands of jobs)

	1990	1991	1992	1993	1994	1995
Chemicals and allied products	484.0	477.0	479.0	474.0	471.0	467.0
Inorganic chemicals	49.5	51.1	49.4	44.1	41.1	n/a
Petrochemicals	165.0	162.0	158.0	158.0	156.0	n/a
Plastics, resins, and synthetic rubber	37.9	36.7	35.9	36.6	40.6	40.0
Synthetic fibers	43.3	42.7	41.7	39.8	36.6	36.1
Drugs	81.4	82.7	92.5	94.6	102.0	106.0
Soaps, cleaners, and toilet goods	77.8	75.7	74.8	74.6	70.6	69.9
Paints and allied products	27.2	25.2	25.7	25.9	27.0	26.7

Number of Production Workers in Various Fields of Chemistry and Chemical Engineering, 1990–1995 (in thousands of jobs) *(continued)*

	1990	1991	1992	1993	1994	1995
Organic chemicals	72.7	72.5	70.6	70.9	65.7	67.0
Fertilizers, nitrogen	4.8	4.7	4.7	4.7	5.4	5.3
Fertilizers, phosphatic	7.5	7.3	6.7	6.6	6.2	6.1
Agricultural chemicals	10.2	9.1	9.6	9.2	8.8	8.7
Adhesives and sealants	11.9	11.7	11.6	11.5	11.3	11.2
Petroleum refining	47.3	47.6	47.9	47.3	46.3	n/a

Source: U.S. Department of Commerce, Bureau of the Census, *Statistical Abstract of the United States, 1996*, Tables 1403–15.

Average Hourly Earnings in Various Fields of Chemistry and Chemical Engineering, 1990–1995 (in dollars)

	1990	1991	1992	1993	1994	1995
Chemicals and allied products	14.40	14.98	15.27	15.62	16.21	16.93
Inorganic chemicals	15.75	16.14	16.59	16.90	17.64	n/a
Petrochemicals	16.50	17.05	17.63	17.93	18.57	n/a
Plastics, resins, and synthetic rubber	16.88	17.25	18.71	18.75	19.37	20.24
Synthetic fibers	13.62	14.46	14.22	14.63	14.04	14.65
Drugs	14.22	15.36	14.76	15.80	16.25	17.36
Soaps, cleaners, and toilet goods	11.73	12.25	12.32	12.26	12.37	12.93
Paints and allied products	11.89	12.09	12.72	13.00	13.98	14.61
Organic chemicals	17.73	18.29	18.70	18.88	20.23	n/a
Fertilizers, nitrogen	13.80	14.77	15.65	16.11	16.69	17.45
Fertilizers, phosphatic	13.70	14.65	13.82	14.70	16.69	17.45
Agricultural chemicals	16.27	15.15	15.59	16.41	18.40	19.23
Adhesives and sealants	11.50	11.43	12.41	12.57	13.41	14.02
Petroleum refining	18.56	19.46	19.74	21.04	21.64	n/a

Source: U.S. Department of Commerce, Bureau of the Census, *Statistical Abstract of the United States, 1996*, Tables 1403–15.

Value of Imports of Chemical Products into the United States, 1990–1995 (in millions of dollars)

	1990	1991	1992	1993	1994	1995
Chemicals and allied products	22,173	23,808	27,213	28,502	33,407	40,260
Inorganic chemicals	4,324	4,284	4,166	4,039	4,707	5,607
Petrochemicals	9,667	9,941	11,084	12,099	14,599	17,671
Plastics, resins, and synthetic rubber	1,811	1,776	2,063	2,520	3,281	4,086
Synthetic fibers	703	784	907	1,127	1,302	1,384
Drugs	2,303	2,854	3,277	3,101	3,366	4,632
Soaps, cleaners, and toilet goods	1,000	1,076	1,308	1,463	1,689	1,936

Value of Imports of Chemical Products into the United States, 1990–1995 (in millions of dollars) (continued)

	1990	1991	1992	1993	1994	1995
Paints and allied products	137	141	184	197	261	343
Organic chemicals	6,272	6,583	7,242	7,330	8,654	10,566
Adhesives and sealants	90	94	112	119	135	140
Petroleum refining	14,269	10,890	10,185	9,742	9,274	8,765

Source: U.S. Department of Commerce, Bureau of the Census, *Statistical Abstract of the United States*, 1996, Tables 1403–15.

Value of Exports of Chemical Products from the United States, 1990–1995 (in millions of dollars)

	1990	1991	1992	1993	1994	1995
Chemicals and allied products	38,989	42,974	43,964	45,065	51,599	60,761
Inorganic chemicals	4,662	4,990	5,130	4,822	5,215	6,178
Petrochemicals	21,173	23,103	22,150	22,047	26,555	32,763
Plastics, resins, and synthetic rubber	6,316	7,447	7,052	7,225	8,478	10,350
Synthetic fibers	1,627	1,594	1,473	1,407	1,594	2,078
Drugs	5,050	5,731	6,774	7,222	7,565	7,996
Soaps, cleaners, and toilet goods	1,853	2,235	2,529	2,853	3,358	3,710
Paints and allied products	570	686	754	811	936	1,009
Organic chemicals	9,738	10,063	10,036	10,409	12,336	15,606
Adhesives and sealants	159	169	202	241	283	315
Petroleum refining	5,988	6,356	5,759	5,641	4,958	5,333

Source: U.S. Department of Commerce, Bureau of the Census, *Statistical Abstract of the United States*, 1996, Tables 1403–15.

Jobless Rates for Chemists, 1991–1997

Year	1991	1992	1993	1994	1995	1996	1997
Percent Unemployment	1.6%	1.9%	2.0%	2.7%	2.6%	3.0%	2.0%

Source: Michael Heylin, "Job Situation for Chemists Takes Turn for the Better," *Chemical & Engineering News*, 28 July 1997, 14.

Median Salaries for Chemists, 1997 (in thousands of dollars)

All chemists	$63.0
By degree	
Ph.D.	$71.0
M.S.	$56.2
B.S.	$49.4
By employer	
Industry	$67.0
Government	$61.8
Academia	$54.0
By gender	
Female	$50.2
Male	$66.6

Source: Michael Heylin, "Job Situation for Chemists Takes Turn for the Better," *Chemical & Engineering News*, 28 July 1997, 15.

Median Salaries for Chemists, 1992–1997 (in thousands of dollars)

	1992	1993	1994	1995	1996	1997
All chemists	54.6	56.0	57.9	59.7	60.0	63.0
Ph.D.	60.0	62.8	65.0	66.0	68.0	71.0
M.S.	50.0	51.5	52.0	53.5	53.6	56.2
B.S.	42.0	43.5	44.3	45.4	45.0	49.4

Source: Reprinted in part with permission from Michael Heylin, "Job Situation for Chemists Takes Turn for the Better," *Chemical & Engineering News*, 28 July 1997, 16. Copyright 1997, American Chemical Society.

Median Salaries for Chemists, by Area of Employment within Industry, 1997 (in thousands of dollars)

	Ph.D.	M.S.	B.S.
General Management	98.0	67.0	63.1
R & D Management	96.5	74.8	68.5
Basic Research	73.4	54.0	43.0
Applied Research	72.2	56.1	50.0
Production	71.0	54.0	46.0
Marketing	81.0	66.0	54.7

Source: Michael Heylin, "Job Situation for Chemists Takes Turn for the Better," *Chemical & Engineering News*, 28 July 1997, 18.

Median Salaries for Chemists, by Area of the U.S., 1997 (in thousands of dollars)

	Ph.D.	M.S.	B.S.
Pacific	70.0	57.3	51.2
Mountain	66.2	51.6	47.2
West North Central	67.9	54.1	48.0
West South Central	68.7	54.7	49.8
East North Central	72.0	57.6	47.8
East South Central	60.6	51.0	47.6
Northeast	74.0	54.0	48.0
Middle Atlantic	75.0	57.0	52.0
South Atlantic	70.0	56.4	50.0

Source: Michael Heylin, "Job Situation for Chemists Takes Turn for the Better," *Chemical & Engineering News*, 28 July 1997, 19.

LARGEST CHEMICAL COMPANIES, UNITED STATES AND WORLDWIDE

Top 50 U.S. Chemical Industries, 1997 (chemical sales in millions of dollars)

Rank	Company	Chemical Sales (1996)
1	Dow Chemical	$18,988.0
2	DuPont	$18,044.0
3	Exxon	$11,430.0
4	Monsanto	$7,267.0
5	Hoechst Celanese	$6,909.0
6	General Electric	$6,487.0
7	Union Carbide	$6,106.0
8	Amoco	$5,698.0
9	Eastman Chemical	$4,782.0
10	BASF Corp.	$4,707.0
11	Huntsman	$4,500.0
12	Occidental Petroleum	$4,484.0
13	Praxair	$4,449.0
14	Shell Oil	$4,305.0
15	Allied Signal	$4,013.0
16	ICI Americas	$4,000.0
17	Rohm and Haas	$3,982.0
18	Arco Chemical	$3,775.0
19	Ashland Oil	$3,695.0
20	Air Products	$3,672.0
21	W. R. Grace	$3,454.1
22	Chevron	$3,422.0
23	Phillips Petroleum	$3,185.0
24	Millennium	$3,040.0
25	Mobil	$3,023.0
26	IMC Global	$2,981.0
27	Dow Corning	$2,532.3
28	National Starch	$2,455.4
29	Lyondell Petrochemical	$2,411.0
30	Akzo Nobel, Inc.	$2,372.9
31	Rhône-Poulenc, Inc.	$2,300.0
32	FMC	$2,293.1
33	Great Lakes Chemical	$2,211.7
34	Hercules	$2,060.0
35	Solvay America	$2,000.0
36	Witco	$1,884.4
37	Elf Atochem	$1,800.0
38	PPG Industries	$1,612.0
39	Morton International	$1,611.7
40	Lubrizol	$1,597.6
41	Olin	$1,586.0
42	Crompton & Knowles	$1,519.1
43	CF Industries	$1,468.2
44	International Flavors	$1,436.1
45	Cabot	$1,434.3
46	Zeneca, Inc.	$1,400.0
47	Engelhard	$1,367.0
48	Farmland Industries	$1,336.3
49	Arcadian Partners	$1,314.0
50	Nalco Chemical	$1,303.5

Source: Reprinted in part with permission from "Finances: Sales, Earnings Drop," *Chemical & Engineering News*, 23 June 1997, 48–49. Copyright 1997, American Chemical Society.

Top 50 Chemical Industries, Worldwide, 1997 (chemical sales in millions of dollars)

Rank	Company	Chemical Sales (1996)
1	BASF (Germany)	$26,519.3
2	Hoechst (Germany)	$19,545.3
3	Bayer (Germany)	$19,543.3
4	Dow Chemical (U.S.)	$18,988.0
5	DuPont (U.S.)	$18,044.0
6	Shell (U.K./Netherlands)	$14,631.3
7	Novartis (Switzerland)	$13,111.3
8	ICI (U.K.)	$12,629.7
9	Exxon (U.S.)	$11,430.0
10	Elf Aquitaine (France)	$10,501.0
11	Rhône-Poulenc (France)	$9,117.2
12	Dainippon Ink & Chemicals (Japan)	$8,696.2
13	Toray Industries (Japan)	$7,860.4
14	Mitsubishi Chemical (Japan)	$7,797.7
15	Monsanto (U.S.)	$7,267.0
16	Veba (Germany)	$7,028.8
17	Sumitomo (Japan)	$6,974.6
18	Norsk Hydro (Norway)	$6,725.6
19	ENI (Italy)	$6,628.6
20	Akzo Nobel (Netherlands)	$6,560.9
21	General Electric (U.S.)	$6,487.0
22	Union Carbide (U.S.)	$6,106.0
23	Henkel (Germany)	$6,099.3
24	Formosa Plastics Group (Taiwan)	$6,079.2
25	Solvay (Belgium)	$5,897.9
26	DSM (Netherlands)	$5,853.8
27	Amoco (U.S.)	$5,698.0
28	Air Liquide (France)	$5,640.8
29	SABIC (Saudi Arabia)	$5,550.0
30	British Petroleum (U.K.)	$5,170.3
31	Asahi Chemical (Japan)	$5,034.0
32	Showa Denko (Japan)	$4,966.7
33	Total (France)	$4,807.2
34	Eastman Chemical (U.S.)	$4,782.0
35	Unilever (U.K./Netherlands)	$4,571.6
36	Huntsman (U.S.)	$4,500.0
37	Occidental Petroleum (U.S.)	$4,484.0
38	Praxair (U.S.)	$4,449.0
39	BOC (U.K.)	$4,397.2
40	Degussa (Germany)	$4,362.7
41	Sekusui Chemical (Japan)	$4,284.1
42	Zeneca (U.K.)	$4,232.8
43	Allied Signal (U.S.)	$4,013.0
44	Rohm and Haas (U.S.)	$3,982.0
45	Roche (Switzerland)	$3,825.0
46	Arco Chemical (U.S.)	$3,775.0
47	Ashland Oil (U.S.)	$3,695.0
48	Mitsui Petrochemical (Japan)	$3,673.0
49	Air Products (U.S.)	$3,672.0
50	Mitsui Toatsu (Japan)	$3,663.2

Source: Reprinted in part with permission from Patricia L. Layman, "Slowdown for Global Top 50," *Chemical & Engineering News*, 21 July 1997, 16–17. Copyright 1997, American Chemical Society.

TOXIC CHEMICAL RELEASES AND BUDGET ALLOCATIONS

Toxic Chemical Release, 1997 (in millions of pounds)

Compound	Amount Released
Methanol	245.0
Ammonia	195.1
Toluene	145.9
Nitrates	137.7
Xylenes	95.7
Zinc compounds	87.6
Hydrochloric acid	85.3
Carbon disulfide	84.2
n-Hexane	77.4
Methyl ethyl ketone	70.1
Chlorine	66.3
Phosphoric acid	57.6
Dichloromethane	57.3
Manganese compounds	45.0
Glycol ethers	43.9
Copper compounds	43.6
Styrene	41.9
Ethylene	34.1
Acetonitrile	28.9
n-Butyl alcohol	27.7

Source: U.S. Environmental Protection Agency

Chemistry-Related U.S. Federal Budget Allocations, 1998 Budget (in millions of dollars)

Item	1996	1997	Actual 1998	1999	Estimate 2000	2001	2002
I. General Science, Space, and Technology							
A. National Science Foundation							
Discretionary	3,156	3,207	3,305	3,311	3,318	3,325	3,332
Mandatory	24	38	38	31	31	31	31
B. Department of Energy							
General programs[D]	966	996	1,003	996	996	996	996
C. NASA							
Space flight, aeronautics, and technology[D]	5,032	4,746	4,722	4,789	4,947	5,135	5,172
Human space flight[D]	5,457	5,540	5,327	5,306	5,077	4,832	4,676
Mission support[D]	2,065	2,123	2,064	2,006	1,889	1,928	2,031
Other NASA programs[D]	16	17	18	19	19	19	19
Total for General Science, Space, and Technology	16,716	16,667	16,477	16,458	16,277	16,266	16,257
II. Energy							
R & D[D]	3,323	3,094	3,062	3,400	3,210	3,085	3,011
Naval petroleum reserves[D]	148	144	117	48	48	48	48
Naval petroleum reserves[M]	-419	-444	-175	-20	-10	-10	-10
Uranium enrichment activities[D]	343	201	249	220	190	190	190
Nuclear waste program[D]	151	182	190	190	190	190	190
Energy conservation[D]	533	550	688	691	688	690	689
Emergency energy preparedness[D]	—	—	209	209	209	209	209
Nuclear Regulatory Commission[D]	18	15	19	19	19	19	19

Chemistry-Related U.S. Federal Budget Allocations, 1998 Budget (in millions of dollars) *(continued)*

Item	1996	1997	Actual 1998	1999	Estimate 2000	2001	2002
Total for Energy (all programs)	2,647	961	1,890	1,172	2,079	1,344	-107
III. Pollution Control and Abatement[D]							
Total	6,635	6,900	7,753	7,847	7,245	7,262	7,369
Superfund and other mandatory programs	-205	-147	-125	-100	-100	-101	-101
IV. National Oceanic and Atmospheric Administration[D]							
Total	1,933	1,977	2,052	2,256	2,129	2,066	2,008

Source: *Budget of the United States Government, Fiscal Year 1988*. Washington, DC: Government Printing Office, 1997, 220–25.
D = Discretionary spending
M = Mandate spending

TRENDS IN PRODUCTION, IMPORTATION, EXPORTATION, AND USE OF CERTAIN CHEMICAL RAW MATERIALS

Production of Mineral Elements, 1992–1996[1]

Element	1992	1993	1994	1995	1996[2]
Aluminum[a]					
primary[3]	4,042	3,695	3,299	3,375	3,600
secondary[4]	1,730	2,540	3,380	2,970	2,800
Antimony[b]					
primary	20,100	22,000	25,500	23,500	22,700
secondary	n/a	n/a	n/a	n/a	n/a
Arsenic[b]	740	767	1,330	557	250
Barium (as barite)[a]	326	315	583	543	650
Beryllium[b]	193	198	173	202	217
Boron[a]	554	574	550	495	622
Bromine[a]	171	177	195	218	227
Cadmium[b]	1,620	1,090	1,010	1,270	1,450
Chromium (secondary)[a]	102	92	99	113	104
Cobalt (secondary)[b]	1,610	1,570	1,570	1,540	1,500
Copper[a]	1,760	1,800	1,850	1,850	1,900
Germanium[c]	13,000	10,000	10,000	10,000	18,000
Gold[b]					
mined	330	331	327	320	325
refined, primary	284	243	241	n/a	n/a
refined, secondary	163	152	148	n/a	n/a
Helium[d]	92.0	99.3	112.0	101.0	103.2
Iodine[e]	2,000	1,940	1,630	1,220	1,200
Iron ore[f]	55.6	55.7	58.4	62.5	60.0
Lead[a]					
mined	407	362	370	394	430
refined, primary	305	335	351	374	340
refined, secondary	861	838	877	926	970
Magnesium[a]	418	386	345	360	450
Mercury[b]					
secondary industrial sources	176	350	466	534	550

Production of Mineral Elements, 1992–1996[1] (continued)

Element	1992	1993	1994	1995	1996[1]
Molybdenum[b]	49,700	36,800	46,800	60,900	57,000
Nickel[a]					
mine	6,670	2,460	—	1,560	—
plant	8,960	4,880	—	8,290	14,600
Palladium[c]	5,440	6,780	6,440	5,260	5,000
Platinum[c]	1,650	2,050	1,960	1,590	1,600
Rare earths[b]	20,700	17,800	20,700	22,200	20,000
(bastnasite concentrates only; other sources withheld.)					
Rhenium[c]	16,000	12,200	15,500	17,000	18,500
Selenium[b]	243	283	360	373	350
Silicon[a]	370	367	390	396	414
Silver[b]					
mined	1,800	1,640	1,490	1,640	1,800
refined, primary[3]	2,160	1,790	1,810	—[g]	—[g]
refined, secondary[4]	1,760	2,020	1,700	—[g]	—[g]
Sulfur[a] (all forms)	10,700	11,100	11,500	11,800	11,800
Tin[b] (secondary)	13,700	12,000	11,700	11,100	12,000
Vanadium[b5]	1,350	2,870	2,830	1,980	2,650
mined	523	488	570	614	620
primary	272	240	217	232	230
secondary	128	141	139	131	130

[a] In thousands of metric tons
[b] In metric tons
[c] In kilograms
[d] In millions of cubic meters
[e] In thousands of kilograms
[f] In millions of metric tons
[g] Data under review
[1] U.S. production figures for certain metals are withheld to protect the industries involved.
[2] Estimated
[3] *Primary* means from mines.
[4] *Secondary* means from scrap.
[5] from petroleum residues only information withheld for other sources.
Source: *Minerals Commodity Summaries*, http://minerals.er.usgs.gov/minerals/pubs/mcs/

Importation of Mineral Elements, 1992–1996

Element	1992	1993	1994	1995	1996[1]
Aluminum[a]	1,730	2,540	3,380	2,970	2,800
Antimony[b]	31,200	30,900	41,500	36,600	35,000
Arsenic[b]					
metal	740	767	1,330	557	250
as As_2O_3	30,700	27,500	26,800	29,000	29,000
Barium (as barite)[a]	366	815	1,081	1,050	1,370
Beryllium[b]	6	8	53	32	45
Boron[a] (all forms)	94	296	176	223	220
Bromine[a]	15	19	24	12	12
Bismuth[b]	1,620	1,330	1,660	1,450	1,600
Cadmium[b]	1,960	1,420	1,110	848	720
Chromium[a]	324	330	273	416	400
Cobalt[b]	5,760	5,950	6,780	6,440	6,900
Copper[a]	593	637	763	808	950
Gallium[c]	8,480	15,600	16,900	18,100	25,000

Importation of Mineral Elements, 1992–1996 *(continued)*

Element	1992	1993	1994	1995	1996[1]
Germanium[c]	13,000	15,000	15,000	16,000	25,000
Gold[b]	159	144	114	126	160
Hafnium[b]	2	3	5	5	7
Indium[b]	36.3	73.4	70.2	85.2	45.0
Iodine[e]	3,750	3,620	4,360	3,950	4,000
Iridium[c]	207	896	926	1,450	2,000
Iron ore[f]	12.5	14.1	17.5	17.5	17.0
Lead[a]	5	1	1	3	5
Lithium[b]	770	810	851	1,140	1,200
Magnesium[a]	179	256	287	328	250
Manganese[a] (all forms)	808	895	940	1,009	1,090
Mercury[b]	92	40	129	377	550
Molybdenum[b]	3,831	3,400	2,280	5,570	5,000
Nickel[a]					
ore	3,580	2,970	—	8,200	16,800
primary	119,000	126,000	127,000	149,000	146,000
secondary	9,510	6,710	6,060	7,900	7,660
Niobium[2][a]	2,450	2,190	2,590	3,580	2,900
Osmium[c]	57	130	55	73	600
Palladium[c]	61,100	78,900	92,500	124,000	110,000
Platinum[c]	57,600	57,200	56,500	71,500	70,000
Rare earths[b]					
metals and alloys	352	235	284	905	442
thorium ore	—	—	—	22	—
cerium compounds	806	1,270	1,890	4,091	4,723
mixed rare-earth oxides	295	249	354	678	918
rare-earth chlorides	728	1,080	2,410	1,249	988
rare-earth oxides	3,100	3,730	5,140	6,499	13,669
ferrocerium alloys	94	105	92	78	97
Rhenium[c]	12,100	5,900	8,200	12,800	15,500
Rhodium[c]	7,750	7,210	7,820	9,600	10,000
Ruthenium[c]	2,740	4,490	9,880	7,520	20,000
Selenium[b]	371	382	411	324	350
Silicon[a]	193	212	255	250	226
Silver[b]	3,220	2,500	2,600	3,250	3,000
Strontium[b]	32,700	26,900	36,000	33,500	35,000
Sulfur[a]	2,730	2,040	1,650	2,510	2,000
Tellurium[b]	48	45	27	46	70
Thallium[c]	838	273	630	1,179	200
Thorium[b] (all forms)	14	18	3	60	41
Tin[b]	27,300	33,700	32,400	33,200	33,000
Titanium[b]	684	2,160	6,470	7,560	8,970
Tungsten[b]	2,500	1,700	3,000	4,200	3,100
Vanadium[b] (all forms)	1,789	3,188	4,145	4,511	4,225
Zinc[a] (all forms)	689	757	820	866	850
Zirconium[b] (all forms)	24,300	45,500	53,300	60,800	61,900

[a] In thousands of metric tons
[b] In metric tons
[c] In kilograms
[d] In millions of cubic meters
[e] In thousands of kilograms
[f] In millions of metric tons
[1] Estimated
[2] In the form of a steel product by the name of ferrocolumbium
Source: *Minerals Commodity Summaries*, http://minerals.er.usgs.gov/minerals/pubs/mcs/

Production of Non-Elemental Minerals, 1992-1996

Mineral	1992	1993	1994	1995	1996[1]
Asbestos[a]	16.0	14.0	10.0	9.0	9.0
Diatomite[a]	595.0	599.0	613.0	687.0	700.0
Feldspar[a]	725.0	770.0	767.0	880.0	900.0
Fluorspar[a] (all forms)	158.0	176.0	146.0	149.0	102.0
Gemstones[g]					
natural	66.2	57.7	50.5	60.0	62.0
synthetic	18.9	18.1	22.2	26.0	26.0
Gypsum[a] (all forms)	30,530.0	31,846.0	34,850.0	34,820.0	35,500.0
Mica, sheet[b]	h	h	h	h	h
Mica, scrap and flake[a]					
mine	85.0	88.0	110.0	108.0	109.0
ground	84.0	92.0	95.0	98.0	100.0
Potash[a] (as K_2O)	1,710.0	1,510.0	1,400.0	1,480.0	1,380.0
Pumice[a]	481.0	469.0	490.0	529.0	465.0
Salt[a]	36,000.0	39,200.0	40,100.0	42,100.0	40,100.0
Soda ash[a]	9,380.0	8,960.0	9,320.0	10,100.0	10,100.0
Talc and pyrophyllite[a]	997.0	968.0	935.0	1,060.0	976.0
Vermiculite[a]	190.0	190.0	180.0	170.0	—⌐

[a] In thousands of metric tons
[b] In metric tons
[c] In kilograms
[d] In millions of cubic meters
[e] In thousands of kilograms
[f] In millions of metric tons
[g] In millions of dollars
[h] Information withheld to protect company business interests
[i] withheld
[1] Estimated
Source: *Minerals Commodity Summaries*, http://minerals.er.usgs.gov/minerals/pubs/mcs/

Importation of Non-Elemental Minerals, 1992–1996

Mineral	1992	1993	1994	1995	1996[1]
Asbestos[a]	32	31	26	22	23
Diatomite[a]	h	h	h	h	h
Feldspar[a]	13	7	7	9	10
Fluorspar[a] (all forms)	640	596	601	672	689
Gemstones[g]	4,950	5,850	6,440	6,540	7,140
Gypsum[a]	7,180	7,390	8,470	8,160	8,000
Mica, sheet[b]	3,460	4,310	2,610	4,230	6,190
Mica, scrap and flake[a]	12	14	18	16	18
Potash[a] (as K_2O)	4,250	4,360	4,800	4,820	4,850
Pumice[a]	257	143	143	238	200
Salt[a]	5,390	5,870	9,630	7,090	9,500
Soda ash[a]	72	89	79	83	95
Talc and pyrophyllite[a]	80	100	155	146	142
Vermiculite[a]	40	30	30	30	30

[a] In thousands of metric tons
[b] In metric tons
[c] In kilograms
[d] In millions of cubic meters
[e] In thousands of kilograms
[f] In millions of metric tons
[g] In millions of dollars
[h] Less than half of a unit
[1] Estimated
Source: *Minerals Commodity Summaries*, http://minerals.er.usgs.gov/minerals/pubs/mcs/

PRODUCTION, IMPORTATION, EXPORTATION, AND CONSUMPTION OF FOSSIL FUELS

Coal Production, 1992–1996 (in thousands of short tons)

Region and State	1992	1993	1994	1995	1996[1]
Appalachian Total	456.6	409.7	445.4	434.9	445.1
Alabama	25.8	24.8	23.3	24.6	24.3
Kentucky, Eastern	119.4	120.2	124.4	118.5	114.9
Maryland	3.3	3.4	3.6	3.7	3.9
Ohio	30.4	28.8	29.9	26.1	28.5
Pennsylvania Total	69.0	59.7	62.2	61.6	68.3
Anthracite	3.5	4.3	4.6	4.7	4.6
Bituminous	65.5	55.4	57.6	56.9	63.8
Tennessee	3.5	3.0	3.0	3.2	3.6
Virginia	43.0	39.3	37.1	34.1	35.7
West Virginia Total	162.2	130.5	161.8	163.0	165.7
Northern	50.0	33.8	49.3	46.1	44.9
Southern	112.1	96.7	112.5	116.9	120.8
Interior Total	195.7	167.2	179.9	168.5	172.2
Arkansas	0.1	0.0	0.1	0.0	0.0
Illinois	59.9	41.1	52.8	48.2	46.0
Indiana	30.5	29.3	30.9	26.0	29.8
Iowa	0.3	0.2	0.0	—	
Kansas	0.4	0.3	0.3	0.3	0.2
Kentucky, Western	41.7	36.1	37.2	35.2	35.2
Louisiana	3.2	3.1	3.5	3.7	3.2
Missouri	2.9	0.7	0.8	0.5	0.7
Oklahoma	1.7	1.8	1.9	1.9	1.7
Texas	55.1	54.6	52.3	52.7	55.3
Western Total	345.3	368.5	408.3	429.6	439.5
Alaska	1.5	1.6	1.6	1.7	1.5
Arizona	12.5	12.2	13.1	11.9	10.7
Colorado	19.2	21.9	25.3	25.7	24.2
Montana	38.9	35.9	41.6	39.5	37.9
New Mexico	24.5	28.3	28.0	26.8	24.9
North Dakota	31.7	32.0	32.3	30.1	30.2
Utah	21.3	21.8	24.4	25.2	27.9
Washington	5.3	4.7	4.9	4.9	4.6
Wyoming	190.2	210.1	237.1	263.8	277.8
U.S. TOTAL	997.5	945.4	1,033.5	1,033.0	1056.7

Source: B.D. Hong, "Annual Review 1996: Coal" *Mining Engineering*, May 1997, 43-50 http://www.eia.doe.gov/cneaf/coal/quarterly/1996_review/contents.html

[1] Estimated

Coal Importation and Exportation, 1992–1996 (in thousands of short tons)

Year	Export	Import
1992	102.5	3.8
1993	74.5	7.3
1994	71.4	7.6
1995	88.5	7.2
1996	90.5	7.1

Source: B.D. Hong, "Annual Review 1996: Coal" *Mining Engineering*, May 1997, 43-50 http://www.eia.doe.gov/cneaf/coal/quarterly/1996_review/contents.html

Use of Petroleum Products in the U.S., 1992–1996 (in millions of gallons per day)

Use	1992	1993	1994	1995	1996
Motor gasoline	59.0	57.2	55.0	55.9	57.5
Aviation gasoline	0.2	0.2	0.2	0.2	0.2
Kerosene-type jet fuel	39.8	41.7	45.2	45.7	48.7
Propane (consumer grade)	3.8	3.5	2.2	3.2	3.2
Kerosene					
No. 1 distillate	0.2	0.2	0.4	0.6	0.4
No. 2 distillate	0.5	0.4	0.4	0.3	0.3
Diesel fuel	21.5	20.8	21.3	21.6	21.9
Fuel oil	3.1	2.9	3.4	3.3	3.2
Total	24.6	23.8	24.6	24.9	25.1
No. 4 fuel	0.6	0.6	0.8	0.5	0.4
Residual fuel oil	22.4	17.2	13.5	11.6	12.7

Source: *Petroleum Marketing Monthly*, Table 3: U.S. Refiner Volumes of Petroleum Products to End Users at ftp://ftp.eia.doe.gov/pub/oil-gas/petroleum/data_publications/petroleum_marketing_monthly/current/txt/table03.txt,06/05/1998.

Trends in Natural Gas Production in the U.S., 1992–1996 (in billions of cubic feet)

	1992	1993	1994	1995	1996
Gross withdrawals[a]	22,132.249	22,725.642	23,580.706	23,743.628	24,276.882
Repressuring[b]	2,972.552	3,103.014	3,230.667	3,565.023	3,708.307
Nonhydrocarbon gas removed	280.370	413.971	412.178	388.392	358.535
Vented and flared[c]	167.519	226.743	228.336	283.739	263.048
Marketed production[d]	18,711.808	18,981.915	19,709.525	19,506.474	19,946.992
Extraction loss	871.905	886.455	888.500	907.795	929.530
Total dry gas production[e]	17,839.903	18,095.460	18,821.025	18,598.679	19,017.462

[a] Gas withdrawn from gas and oil wells.
[b] Injection of natural gas into oil and gas formations for pressure maintenance and cycling purposes.
[c] Natural gas released into the air or burned in flares at base site or at processing plants.
[d] Marketed Production = Gross Withdrawals - Repressuring - Nonhydrocarbon Gas Removed - Vented and Flared.
[e] Total Production = Marketed Production - Extraction Loss.

Source: "Summary statistics for natural gas in the United States, 1992-1996," at http://www.eia.doe.gov/fuelnatgas.html, 09/22/97.

CHAPTER NINE
Organizations and Associations

The organizations and associations listed in this chapter are of two distinct types. The first group consists of organizations made up of working chemists; the second group, of associations of companies involved in the chemical industry. Examples of the first group are the American Chemical Society and the American Institute of Chemical Engineers. Examples of the second group are the American Gas Association and the Chemical Specialties Manufacturers Association. These divisions are not rigid, however, as many organizations make provision for inviting both individual and corporate members.

The number of organizations and associations in each group is very large. Space prohibits a complete and detailed listing of all organizations in either group. The choice has been made, instead, to describe in some detail the larger organizations whose members are more directly involved in some aspect of the chemical industry. Then, a selection of smaller organizations and associations, or those less directly involved in the chemical enterprise, is listed at the end of each section.

Organizations and associations *not* included in this listing are invited to submit relevant information for inclusion in the next edition of this book.

ORGANIZATIONS OF CHEMISTS

American Association for Clinical Chemistry, Inc. (AACC)
2101 L Street, NW, Suite 202
Washington, DC 20037-1526
telephone: 202-857-0717
         800-892-1400
fax: 202-887-5093
e-mail: <info@aacc.org>
URL: <http://aacc.org>

The AACC is an international association made up of clinical laboratory professionals, physicians, research scientists, and other individuals concerned with clinical
chemistry. The organization is subdivided into 22 local sections and is governed by an
11-member Board of Directors. The society was founded in 1948 and currently has
about 11,000 members.

AACC members come from 93 different countries, although about 80% are U.S.
citizens. Nearly half of all members are employed in hospital laboratories, another
quarter in independent laboratories, and the remainder in other positions, such as
academic institutions. About half of all AACC members hold Ph.D.s, one-fifth hold
master's degrees, and the rest hold bachelor's degrees.

Association activities fall into five general categories: government affairs, meetings, publications, education, and special features and programs. The purpose of the
government affairs arm of the association is to provide evaluations of pending federal
legislation, regulations, and reports; to develop position statements on important
issues involving clinical chemistry; and to report to AACC members about legislative,
judicial, administrative, and financial issues of interest to them.

The AACC sponsors many scientific meetings each year. Some examples include
the Annual Meeting and Clinical Laboratory Exposition; the Oak Ridge Conference
on Advanced Analytical Concepts for the Clinical Laboratory; the Arnold O. Beckman
Conference, established for the discussion and evaluation of new technologies in
clinical chemistry; the Clinical Chemistry Forum, a meeting at which experts in some
specific field of clinical chemistry present an update on their field; and the San Diego
Conference, devoted to applications of molecular technology to diagnosis and therapy.

In addition to more than 50 books, the AACC regularly publishes 4 periodicals:
Clinical Chemistry, a peer-reviewed journal presenting research reports and other
articles of interest to the profession; *Clinical Laboratory News*, a monthly magazine
that includes news stories, feature articles, and editorial comments on clinical chemistry; *Forensic Urine Drug Testing Newsletter*, a quarterly publication focusing on
laboratory issues involving the testing for illegal drugs; and *AACC News*, a bimonthly
newsletter reporting on section, division, and committee activities.

A major component of the AACC program is education for its members. Included
in this program are such activities and publications as a career education brochure
designed for high school and college students; a postdoctoral training directory;

student research awards; student travel grants; science fair awards; and summer fellowships. The association also makes available to its members a variety of continuing education programs, including self-study courses, audio and video conferences, review courses, and in-service training programs.

American Association for the Advancement of Science (AAAS)
1200 New York Avenue, NW
Washington, DC 20005
telephone: 202-326-6400
e-mail: <webmaster@aaas.org>
URL: <http://www.aaas.org>

The AAAS was founded in 1848 in Philadelphia, making it one of the oldest scientific societies in the United States. Over the years, it has served as the parent organization from which many specialized scientific societies have sprung, including the American Chemical Society, the American Anthropological Association, and the Botanical Society of America.

As defined in the AAAS constitution, the organization's mission is to further the work of scientists, facilitate cooperation among them, foster scientific freedom and responsibility, improve the effectiveness of science in the promotion of human welfare, advance education in science, and increase the public's understanding and appreciation of the promise of scientific methods in human progress.

Membership in the AAAS is open to anyone who wishes to join. Today, more than 143,000 individuals worldwide belong to the association. These individuals include working scientists, science educators, policymakers, and private citizens from every part of the world.

In addition to its individual members, the AAAS has established formal relationships with 285 affiliates from throughout the world. Of these affiliates, 238 are scientific or engineering societies, 44 are state and regional academies of science, and 3 are city academies of science. The combined membership of the affiliated organizations exceeds 10 million individuals.

Work of the association is carried out largely through three major directorates: Education and Human Resources, International, and Science Policy. The association also sponsors 24 sections focused on areas of special interest to members. These sections range from the physical, biological, and health sciences to the social, economic, and applied sciences.

To many people, the AAAS is probably best known for its publication of the weekly journal *Science*. The journal carries reports of recent research, review articles on topics in scientific research, discussions of public policy relating to science and technology, and fascinating editorial and "letters to the editors" columns.

American Chemical Society (ACS)
1155 16th Street, NW
Washington, DC 20036
telephone: 202-872-4600
             800-451-9190

(outside US): 612-520-6798
fax: 612-520-6706
e-mail (membership): <ACSmbr@aol.com>
URL: <http://www.acs.org>

The ACS is the largest single-subject scientific society in the world, with a membership of more than 151,000 members. About 60% of the members are from industry, and the remainder are from academia. Members range from college students to postdoctoral professionals. The organization was founded in 1876 and chartered by the U.S. Congress in 1937.

The ACS mission statement's seven primary objectives are to

- promote the public's perception and understanding of chemistry and the chemical sciences through public outreach programs and public awareness campaigns;
- involve the society's over 151,000 members in improving the public's perception of chemistry;
- assist the federal government with advice on scientific and technological issues involving the chemical sciences;
- enrich professionals in academia and private industry through development programs, peer interactions, and continuing education courses;
- host national, regional and local section meetings for the exchanging of ideas, information, and chemical research discoveries;
- provide career development assistance and employment opportunities for students and professionals in academia and private industry;
- foster communication and understanding between members, the chemical industry, the government, and the community to enhance the quality of scientific research, support economic progress, and ensure public health and safety.

  (As cited in "ACS at a Glance" at <http:// www.acs.org/acsgen/ aboutacs.htm>)

Much of the work of ACS is done through its 34 technical divisions organized around specialized areas of chemical research. These divisions include Agricultural and Food Chemistry; Agrochemicals; *Analytical Chemistry; *Biochemical Technology; *Biological Chemistry; Business Development and Management; *Carbohydrate Chemistry; *Cellulose, Paper and Textile; *Chemical Education; *Chemical Health and Safety; *Chemical Information; *Chemical Technicians; *Chemical Toxicology; *Chemistry and the Law; Colloid and Surface Chemistry; *Computers in Chemistry; *Environmental Chemistry; Fertilizers and Soil Chemistry; Fluorine Chemistry; *Fuel Chemistry; *Geochemistry; *History of Chemistry; *Industrial and Engineering Chemistry; *Inorganic Chemistry; *Medicinal Chemistry; *Nuclear Chemistry and Technology; *Organic Chemistry; Petroleum Chemistry; *Physical Chemistry; *Polymer Chemistry; *Polymeric Materials: Science and Engineering; Professional Relations; *Rubber; and *Small Chemical Businesses. Divisions with an asterisk (*) have their own home pages on the Internet.

The society also maintains 46 committees responsible for a variety of activities. Three of these with broad general interest include the Women Chemists Committee, the Younger Chemists Committee, and the Committee on Minority Affairs. All three committees were formed to increase outreach to people with special needs and interests who would be interested in chemistry.

The ACS Office of International Activities is charged with maintaining contact between the ACS and other scientific societies around the world. Some of the activities supported by this committee include the International Chemistry Celebration, scheduled for April 1999; Project Bookshare, a program through which more than 100 tons of scientific books and journals have been donated to chemists in more than 50 countries; ACS International Initiatives, a program that makes possible visits to U.S. universities by chemists in 14 nations in central and eastern Europe and in Latin America; the Inter-American Young Scientists Exchange Program, which supports research visits in the United States by faculty members from Argentina, Brazil, Chile, and Mexico; the Subcommittee on Scientific Freedom and Human Rights, which monitors complaints about violations of scientific freedom and human rights around the world; and Project Instrument Share, a new program developed along the lines of Bookshare in which chemists in the United States provide instruments to chemical laboratories in other nations throughout the world.

The ACS also has a variety of educational scholarships and grants. For example, Project SEED (Summer Education Experience for the Disadvantaged) makes it possible for economically disadvantaged students to work in academic, industrial, and governmental laboratories for 8- to 10-week periods during the summer. The society also sponsors a number of awards to outstanding chemists in a variety of fields of chemical research and development.

American Institute of Chemical Engineers (AIChE)
345 East 47th Street
New York, NY 10017-2395
telephone: 800-242-4363
fax: 212-705-8400
e-mail: <xpress@aiche.org>
URL: <http://www.aiche.org>

The AIChE was created in Philadelphia on 22 June 1908 when 19 men met to form a new organization for what was, at the time, a new field of study. At the time, no more than about 500 people referred to themselves as "chemical engineers," and there was almost no formal organization of the field. Some experts regard the Philadelphia meeting, in fact, as the birth of the field of chemical engineering.

Today, the AIChE includes more than 50,000 members, from the undergraduate college level to entry-level engineers to chief executives of major corporations. Some fields of specialization included within AIChE membership include petrochemicals, food, pharmaceuticals, textiles, pulp and paper, ceramics, electronic components and chemicals, biotechnology, environmental control and cleanup, and safety engineering.

The organization also consists of 188 local sections, 13 divisions, 26 standing committees, and 148 student chapters. Most inquiries about membership, book purchases, magazine and journal subscriptions, and information about meetings can be obtained from the AIChE*xpress* Service Center at the telephone number or e-mail address listed above.

Among the AIChE's activities are the following:

- *Meetings and Expositions.* The AIChE conducts two major conferences each year, a spring national meeting and an annual meeting. In addition, the organization sponsors many smaller conferences on specialized topics.

- *Education.* The organization conducts more than 200 continuing-education courses on a variety of topics of interest to chemical engineers. They also offer in-house seminars designed for individual companies, academic institutions, and local sections of the association.

- *Sponsored Research.* The AIChE acts as a facilitator in the organization of research projects that individual industrial, academic, and governmental agencies would not be able to carry out on their own. The organization then makes arrangements for the dissemination of research findings obtained from these projects.

- *Member Services.* The organization provides a wide variety of financial services to its members, including health and life insurance plans and savings programs. It also collaborates with companies, academic institutions, governmental agencies, and other bodies in providing pension plans, retirement programs, and other financial services that individual bodies may not be able to provide on their own.

- *Career Guidance.* The AIChE has developed a CD-ROM program and a variety of print and nonprint materials that provide information about a career in chemical engineering.

- *Government Relations.* The organization works at the state and national levels to monitor legislation that will affect chemical engineers and the companies for which they work.

- *Publications.* The AIChE publishes a monthly magazine, *Chemical Engineering Progress*, which discusses state-of-the-art technologies; a monthly newsletter, *AIChExtra*; a peer-reviewed academic journal, the *AIChE Journal*, which reviews recent developments in the field of chemical engineering; and specialized journals, such as *Environmental Progress*, *Process Safety Progress*, and *Biotechnology Progress*. *ChAPTER One* is a special undergraduate version of *Chemical Engineering Progress*, published eight times a year. The organization's *National Engineers Week* deals with AIChE activities and programs.

Among the many books published by the AIChE are *Proceedings of the Ethylene Producers' Conference*; *Heat Transfer*; *Chemical Process Control: Assessment and New Directions for Research*; *Dow's Exposure Index*; *Advances in Fluid Particles and Fluid Particle Systems*; *Ammonia Plant Safety and Related Facilities*; *Advances in Forest Products Industries*; and *Transport Properties and Related Thermodynamics Data of Binary Mixtures*.

International Union of Pure and Applied Chemistry (IUPAC)
P.O. Box 13757
Research Triangle Park, NC 27709-3757
telephone: 919-485-8700
fax: 919-485-8706
e-mail: <secretariat@iupac.org>
URL: <http://www.iupac.org>

The IUPAC was established in 1919 by a group of chemists from around the world who were concerned about the need to set standards in many fields of chemistry. Some of the areas identified as requiring standardization included nomenclature of inorganic and organic compounds, standardization of atomic weights and physical constants, editing of tables of properties of matter, and measures to prevent repetition of identical or similar papers in chemistry. The IUPAC grew out of an earlier group, the International Association of Chemical Societies, which had first met in Paris in 1911 with these objectives in mind.

The IUPAC consists of chemists in two associated groups, the National Adhering Organizations and the Observer Countries. Currently, the IUPAC is composed of 40 adhering organizations and 14 observer countries. Altogether, more than 1,000 chemists throughout the world work with the IUPAC on a variety of assignments on a voluntary basis. The assignments are organized into 32 commissions, which, in turn, are grouped into seven divisions: Physical Chemistry; Inorganic Chemistry; Organic Chemistry; Macromolecular Chemistry; Analytical Chemistry; Chemistry and the Environment; and Human Health.

Today, the IUPAC is the recognized world authority on chemical nomenclature, terminology, symbols, units, atomic weights, and related topics in chemistry. The organization provides advice on chemical matters to various international agencies, such as the World Health Organization; the United Nations Food and Agricultural Organization; the United Nations Education, Scientific, and Cultural Organization; the International Organization for Standardization; and the Organization Internationale de Métrologie Légale.

The IUPAC publishes a wide variety of reports and recommendations dealing with topics such as standards, clinical chemistry practices, spectrochemical and other analytical procedures, and solubility data. It also publishes two journals, *Pure and Applied Chemistry* and *Chemistry International*.

Royal Society of Chemistry
Burlington House
Piccadilly, London W1V 0BN
U.K.
telephone: +44 (0) 171-437-8656
fax: +44 (0) 171-437-8883
e-mail: <waldenn@rsc.org>
URL: <http://www. rsc.org>

The Royal Society of Chemistry is one of the oldest and most prestigious associations of chemists in the world. Its purposes are to provide a forum through which chemists can exchange ideas and research results and to provide a network through which new ventures can be encouraged. Among the activities of the society are the endowment of lectures, the awarding of scholarships and awards, the publication of many journals and books, the maintenance of an extensive library and database system, the promotion of high standards for the training of chemists in the United Kingdom, and the maintenance of a career advice and employment counseling service.

Among the many publications of the Royal Society are *Chemical Communications, Chemical Society Reviews, Journal of Chemical Research, Mendeleev Communications, Russian Chemical Reviews, Analytical Communications, Contemporary Organic Synthesis, Natural Products Reports, Perkin Transactions 1 and 2, Faraday Transactions, Dalton Transactions,* and *Journal of Materials Chemistry.*

Society of Chemical Industry (SCI)
14/15 Belgrave Square
London SW1X 8PS
U.K.
telephone: +44 (0) 171-235-3681
fax: +44 (0) 171-823-1698
e-mail: <info@chemind.demon.co.uk>
URL: <http://sci.mond.org>

The SCI is an international association of individual members founded by Royal Charter in 1881. Its principal goal is to further the application of chemistry and related sciences for the public benefit. About three-quarters of the SCI's members are citizens of the United Kingdom, and the remaining quarter come from a variety of nations around the world. Members include undergraduate and postgraduate students and employees of industrial, academic, governmental, consulting, and contracting agencies. They represent every aspect of chemistry as well as the life sciences; environmental science; materials science; mechanical, electronic, and process engineering; psychology; law; economics; and business administration.

The society maintains 16 geographical sections located throughout the United Kingdom and in the United States and Canada. It also has 20 subject groups concerned with topics such as agriculture and environment, biotechnology, colloid and surface chemistry, fine chemicals, fire chemicals, food engineering, oils and fats, process engineering, sensory and consumer chemistry, and separation science and technology.

The SCI also publishes an extensive list of books, journals, and electronic publications that include the following: the magazine *Chemistry & Industry*; the journals *Journal of Chemical Technology & Biotechnology, Journal of the Science of Food & Agriculture, Pesticide Science,* and *Polymer International*; the events diaries, the *SCI Bulletin, Food Science & Technology Meetings,* and *Environmental Science & Technology Meetings*; books, such as *Nutrition and Sport, Recovery of Bioproducts, Motor Gasoline, Crop Protection Agents from Nature,* and *Antifungal Agents: Discovery and*

Modes of Action; and the electronic publications, *Chemistry Today, Pharmaceuticals Today, Environment Today, Biotechnology Today,* and *Chemistry and Industry Magazine.*

Specialized Organizations and Associations of Chemists

American Association of Cereal Chemists (AACC)
3340 Pilot Knob Road
St. Paul, MN 55121-2097
telephone: 612-454-7250
fax: 612-454-0766
e-mail: <aacc@scisoc.org>
URL: <http://www.scisoc.org/aacc/>

The AACC is an international association of chemists working in the field of cereals and cereal grains. It holds an annual meeting and publishes two journals, *Cereal Chemistry* and *Cereal Foods World*, as well as other books and a textbook, *Principles of Cereal Science and Technology.*

American Association of Pharmaceutical Scientists (AAPS)
1650 King Street, Suite 200
Alexandria, VA 22314-2747
telephone: 703-548-3000
fax: 703-684-7349
e-mail: <aaps@aaps.org>
URL: <http://www.aaps.org>

Founded in 1986, the AAPS has as its mission serving the pharmaceutical sciences, promoting the economic vitality of the pharmaceutical sciences and scientists, and representing scientific interests within academia, industry, government, and other private and public institutions. The purpose of the organization is to serve its members in the pharmaceutical sciences, the biomedical and biotechnological communities, the health professions, and the interest of public health by (1) providing a forum for open interchange and dissemination of scientific knowledge; (2) influencing the formation of public policy relevant to health and related issues of public concern; (3) promoting the pharmaceutical sciences and providing for recognition of individual achievement; and (4) fostering education, career growth, and the personal development of its members.

American Association of Textile Chemists and Colorists (AATCC)
One Davis Drive, P.O. Box 12215
Research Triangle Park, NC 27709-2215
telephone: 919-549-8141
fax: 919-549-8933
e-mail: <info@aatcc.org>
URL: <http://www.aatcc.org>

The AATCC was founded in 1921 and now consists of 7,400 individual and 290 corporate members in the United States and 60 other nations. The association has three primary objectives:

- to promote increased knowledge about the application of dyes and chemicals in the textile industry;
- to encourage research work on chemical processes and materials of importance to the textile industry;
- to provide a channel by which members can exchange professional information with each other.

The association sponsors workshops, symposia, an annual international conference and exhibition, and a variety of research and technology committees. The AATCC publishes a monthly magazine, *Textile Chemist and Colorist*, and the annual *AATCC Technical Manual*.

American Ceramic Society (ACerS)
P. O. Box 6136
Westerville, OH 43086-6136
telephone: 614-890-4700
fax: 614-899-6109
e-mail: <info@acers.org>
URL: <http:// www.acers.org>

The ACerS describes itself as "the world's leading organization dedicated to the advancement of ceramics." It consists of 11 divisions: Basic Science; Cements; Design; Electronics; Engineering Ceramics; Glass and Optical Materials; Materials and Equipment; Nuclear and Environmental Technology; Refractory Ceramics; Structural Clay Products; and Whitewares. The society has more than 30 geographic areas or sections within the United States, including Arizona, Baltimore-Washington, Canton-Alliance, Central Ohio, Central Pennsylvania, Chicago-Milwaukee, Eastern Washington, Hudson-Mohawk Valley, Indiana, Lehigh Valley, Michigan/Northwestern Ohio, Minnesota, New England, New Mexico, New York Metropolitan, and Northern California. It also has more than 30 student branches in colleges and universities throughout the United States.

Each year, more than 1,500 papers are published in the society's journals and books. These include *Ceramic Bulletin, Journal of the American Ceramic Society, Ceramic Abstracts, Ceramic Engineering and Science Proceedings, ceramicSOURCE*, and books in the Ceramic Transactions series. The society sponsors an annual meeting and exposition as well as topical, section, division, and other meetings held throughout the year.

As part of its ongoing program of continuing education, the society offers short courses, a Ceramic Correspondence Institute, and the Ceramic Information Center that offers search services to its members.

American Leather Chemists Association (ALCA)
c/o University of Cincinnati–Tanners Building
P.O. Box 210014
Cincinnati, OH 45221-0014
telephone: 513-556-1197
fax: 513-556-2377
e-mail: <donmezk@email.uc.edu>
URL: <http://ucunix.san.uc.edu/~donmezk/>

The ALCA was founded in 1904 by nine leather chemists for the purpose of establishing approved methods for tanning analysis. The organization has members throughout the United States and Canada and in 32 foreign countries. The association publishes the *Journal of the American Leather Chemists Association* and has established the Stiasny Fellowship at the University of Cincinnati, the John Arthur Wilson Memorial Lecture, the ALCA Physical and Mechanical Properties Committees on Leather, and the ALCA Official Test Methods.

American Oil Chemists' Society (AOCS)
1608 Broadmoor Drive
Champaign, IL 61821-5930
telephone: 217-359-2344
fax: 211-351-8091
e-mail (for membership): <membership@aocs.org>
URL: <http://www.aocs.org>

The AOCS is an international professional association of scientists working in any field related to fats, oils, proteins, surfactants, and detergents. The association has four levels of membership: active, corporate, retired, and student. In addition to holding an annual meeting, the association publishes *INFORM*, a newsletter dealing with business and scientific issues, and two professional journals, *Journal of the American Oil Chemists' Society* and *Lipids*. Student members also receive *Student Common Interest Group Newsletter*.

American Society of Brewing Chemists (ASBC)
3340 Pilot Knob Road
St. Paul, MN 55121-2097
telephone: 612-454-7250
fax: 612-454-0766
e-mail: <asbc@scisoc.org>
URL: <http://www.scisoc.org/asbc/>

The ASBC was founded in 1934 and currently has over 700 individual members and over 50 corporate member firms. The purposes of the organization are to solve technical problems of brewing on an industry-wide basis, to keep current on technical needs of the brewing industry, and to anticipate future concerns in the industry.

AOAC INTERNATIONAL
481 North Frederick Avenue, Suite 500
Gaithersburg, MD 20877-2417
telephone: 301-924-7077
              800-379-2622
fax: 301-924-7087
e-mail: <aoac@aoac.org>
URL: <http://www.aoac.org>
AOAC INTERNATIONAL was established in 1884 as the Association of Official
Analytical Chemists. The organization was created to promote the validation of
methods used in analysis and to promote quality measurements in the analytical
sciences. Today, more than 3,800 analytical scientists from industry, government,
and academia in over 90 countries are members of AOAC. These professionals come
from the fields of analytical chemistry, microbiology, biochemistry, toxicology,
spectroscopy, and forensic science. They hold positions in management, administration, and the laboratory.

A major focus of AOAC activities is its three validation programs: the AOAC®
Official Methods Program, the AOAC® Peer-Verified Methods Program, and the
AOAC® Performance-Tested Test Kit Program. The purpose of these programs is to
help analytical scientists make informed choices about the appropriateness of particular methods of analysis.

Publications of the association include a monthly newsletter, *Inside Laboratory
Management*; a journal, *Journal of AOAC INTERNATIONAL*; and the "bible" of the
association, *Official Methods of Analysis of AOAC INTERNATIONAL*.

The organization has 12 sections and 2 subsections in the United States, Canada,
Europe and non-European Mediterranean countries, South and Central America,
Mexico, and the Caribbean.

The Electrochemical Society (ECS)
10 South Main Street
Pennington, NJ 08534-2896
telephone: 609-737-1902
fax: 609-737-2743
e-mail: <ecs@electrochem.org>
URL: <http://www.electrochem.org>
The ECS is an international, nonprofit, educational organization concerned with a
broad range of phenomena relating to electrochemical and solid-state science and
technology. Two major publications of the society include the monthly *Journal of the
Electrochemical Society*, containing more than 60 research papers, and the quarterly
Electrochemical Society Interface, which includes both scientific articles and information about individuals, programs, and activities of the society. The society also
publishes Proceedings Volumes, Meetings Abstract Volumes, and Monograph Volumes on a broad range of topics in the field of electrochemistry and solid-state
physics.

The Geochemical Society (GS)
287 Scott Hall
1090 Carmack Road
Columbus, OH 43210
telephone: 614-292-6919
fax: 614-292-7273
e-mail: <geochsoc@magnus.acs.ohio-state.edu>
URL: <http://www.ciw.edu/geochemical_society/>
The Geochemical Society consists of scientists interested in the application of chemistry to the solution of geological and cosmological problems. Members of the society deal with a wide range of issues, including physical chemistry, igneous and metamorphic petrology, chemical processes in natural systems, organic geochemistry, meteoritics, lunar science, geochemistry of the solar system, and cosmochemistry. The society publishes a bimonthly research journal, *Geochimica et Cosmochimica Acta*; a biannual newsletter, *The Geochemical News*; and a series of special publications consisting of original papers written to honor important geochemists.

Materials Research Society (MRS)
9800 McKnight Road
Pittsburgh, PA 15237-6006
telephone: 412-367-3003
fax: 412-367-4373
e-mail: <info@mrs.org>
URL: <http://www.mrs.org>

The MRS is a nonprofit organization of scientists, engineers, and research managers from government, industry, academia, and research laboratories that was founded for the purposes of sharing new information on research and the development of new materials of technological importance. The organization was established in 1973 and now consists of more than 12,000 members, from the United States and about 50 other countries throughout the world.

The organization sponsors 2 annual meetings, which include about 60 symposia on specialized topics. It also includes student university chapters and regional sections.

In addition to the publication of symposia proceedings, the MRS publishes the *MRS Bulletin*, the *Journal of Materials Research*, and a variety of other print materials, databases, and videotapes relating to current research in materials development.

Society of Cosmetic Chemists (SCC)
120 Wall Street, Suite 2400
New York, NY 10005-4088
telephone: 212-688-1500
fax: 212-668-1504
e-mail: <societycoschem@worldnet.att.net>
URL: <http://www.scconline.org>

The SCC was founded in 1945 for the purpose of promoting high standards of practice in the cosmetic sciences. The association attempts to increase and disseminate scientific knowledge about cosmetics through meetings and publications; to promote research in cosmetic science and industry; and to set high ethical, professional, and educational standards. Its mission statement says that its purpose is "to further the interests and recognition of cosmetic scientists while maintaining the confidence of the public in the cosmetic and toiletries industry." The SCC currently has over 3,500 members.

Society of Environmental Toxicology and Chemistry (SETAC)
1010 North 12th Avenue
Pensacola, FL 32501-3370
telephone: 904-469-1500
fax: 904-469-9778
e-mail: <setac@setac.org>
URL: <http://www.setac.org>

The SETAC was founded in 1979 to provide a forum through which individuals from many disciplines interested in environmental issues could meet and interact. Today, as in the past, the organization consists of an almost equal mix of members from academia, business, and government. The SETAC consists of more than 5,000 members, representing all 50 states and 55 other nations throughout the world. The organization publishes the monthly journal *Environmental Toxicology and Chemistry*.

ORGANIZATIONS OF CHEMICAL COMPANIES

Chemical Manufacturers Association (CMA)
1300 Wilson Boulevard
Arlington, VA 22209
telephone: 703-741-5000
fax: 703-741-6000
e-mail: <webmaster@mail.cmahq.com>
URL: <http://www.cmahq.com>

The CMA was founded in 1872 and lists more than 200 members and partners. Together, the members and partners of the CMA are responsible for the production of about 90% of the basic industrial chemicals in the United States. The association sponsors a wide variety of programs and activities in which more than 2,000 scientists, engineers, and health, safety, and environmental managers from member companies participate.

One of the association's most important programs is called Responsible Care®, established in 1988. The purposes of Responsible Care® are to help member companies and partners to "improve performance in health, safety and environmental quality, to listen and respond to public concerns, to assist each other to achieve optimum performance, and to report their progress to the public."

The basis of Responsible Care® is a set of six codes of management practices that deal with community awareness and emergency response, pollution prevention, process safety, distribution, employee health and safety, and product stewardship.

Chemical Specialties Manufacturers Association (CSMA)
1913 Eye Street, NW
Washington, DC 20006
telephone: 202-872-8110
fax: 202-872-8114
e-mail: <CSMA@Juno.com>
URL: <http://www.csma.org>

The CSMA was founded in 1914 and currently represents over 400 companies involved in the manufacture, formulation, distribution, and sale of chemical specialty products for household, institutional, and industrial use. The organization's mission statement defines the following objectives:

- to foster high standards for the industry and concern for the health, safety and environmental impacts of its products;
- to address legislative and regulatory challenges at the federal, state, and local levels;
- to meet the needs of industry for technical and legal guidance;
- to provide a forum to share ideas for scientific and marketing excellence.

The CSMA sponsors two major annual conventions each year, the Annual Meeting in December and the Mid-Year Meeting in May. The latter meeting also includes the annual CHEM-MART trade show. In addition to these meetings, the organization sponsors seminars on six topics of concern to the industry: aerosols, antimicrobials, detergents, pesticides, industrial and automotive specialty products, and polishes and floor-maintenance products.

One of the CSMA's most important functions is its Product Ingredient Review (PIR) program. The purpose of PIR is to bring together producers and marketers of a variety of consumer products, such as pesticides, aerosols, laundry products, and air fresheners, in order to develop product health and safety data. These data are used by the U.S. Environmental Protection Agency and state environmental agencies in approving and registering a wide variety of consumer products.

The CSMA publishes a wide variety of materials including a quarterly journal, *Chemical Times & Trends*, which is devoted to an in-depth study of topics of interest to members of the association; *Executive Newswatch*, a weekly newsletter summarizing legislative, regulatory, and marketing developments; *Vendors to the Trade*, a list of members and vendors; *Consumer Topics in Focus*, a quarterly report on chemical specialties; *Consumer Products Education Bureau Quarterly Report*, a report of developments and trends in the aerosol industry; and a long list of print and nonprint resources.

Included among the latter are consumer information pamphlets, such as "Your Child and Household Safety," "Consumer Products Handbook," "Household Care and Indoor Air Quality," "No Wax Floors: A Mirage or a Miracle?" "Sniffing Abuse: It Can Kill," and "Fact Sheets: Automotive Chemicals."

Among the technical manuals and audiovisual programs produced by the association are "Polishes and Floor Maintenance Products Test Methods Compendium," "Polymer Technology," "Wax Technology," "Floor Safety," "Aerosol Guide," "Safety in the Aerosol Laboratory," and "Aerosol Can Filling Room."

Council for Chemical Research (CCR)
1620 L Street, NW, #825
Washington, DC 20036
telephone: 202-429-3971
fax: 202-429-3976
e-mail: <74534.3471@compuserv.com>
URL: <http://www.ccr.org>

The CCR was established in 1979 as the result of a meeting called by Dr. M. E. Pruitt, Vice President of Research for the Dow Chemical Company. Dr. Pruitt was concerned about the need for greater support for research in the chemical sciences and chemical engineering. The CCR was incorporated in 1980 and moved in 1990 from its original headquarters in Bethlehem, Pennsylvania, to its present location in Washington, DC. The purposes of this move were "to facilitate interactions with the federal government and to coordinate more effectively with other organizations having overlapping interests" (<http://www.chem.purdue.edu/ccr/>).

Members of the CCR include 153 academic institutions, 36 industrial corporations, and 13 governmental laboratories engaged in chemical research or chemical engineering in the United States. Together these institutions have a research-and-development budget of more than $7 billion.

The mission statement of the CCR is "to advance research in chemistry-based sciences, engineering, and technology that benefits society and the national well-being through productive interactions among the industrial, academic, and governmental research sectors."

Among the CCR's activities are an annual meeting at which members have the opportunity to meet and exchange ideas; participation in Speakout Toolbox, a program through which members of CCR, the Chemical Manufacturers Association, the American Chemical Society, and the American Institute of Chemical Engineers talk about the importance of chemistry and chemical technology to adult audiences; partnership with other chemical societies in the Chemical Industry Technology Vision 2020 Project, an effort to develop a strategic vision for the chemical profession in the twenty-first century; sponsorship of conferences on research areas, such as environmental technology, catalysis, polymers, separations, and process analysis; conducting surveys on topics of interest to the chemical community, such as Ph.D. supply and demand; and publication of the "Directory of Industrial and Governmental Laboratory Speakers" for use by academic members of CCR.

Specialized Organizations and Associations of Chemical Companies

Adhesive and Sealant Council, Inc. (ASC)
1627 K Street, NW, Suite 1000
Washington, DC 20006
telephone: 202-452-1500
fax: 202-452-1501
e-mail: <webmaster@listserv.ascouncil.org>
URL: <http://www.ascouncil.org>
The ASC is the primary organization representing the adhesive and sealant industry in North America. Its members are involved in the global development, marketing, distribution, and consumption of adhesives and sealants. Members represent the manufacture, supply, distribution, and end-use of sealants in corporate and individual situations, architecture and engineering, academia, and government.

The Adhesive and Sealant Council Education Foundation (ASCEF) was created in 1987 to create a partnership between industry and the Center for Adhesive and Sealant Science (CASS) at Virginia Tech. Some achievements of the partnership thus far have been an increased ability to attract and train potential employees for the adhesive and sealant industry, improved access to an academic environment and state-of-the-art equipment used in the adhesive and sealant industry, expanded opportunities for continuing education of current and future employees, encouragement of research in the field of adhesives and sealants, and communication and exchange of ideas between industry and academia.

Publications of the society include the newsletter *Catalyst*, also available in electronic format; the *Twin Cities Newsletter*; *The ASC Journal*; and manuals such as the *Hazardous Materials Shipping Manual*, the *VOC/VOS Manual*, and the *Sealant Specifications Compendium*.

American Crop Protection Association (ACPA)
1156 15th Street, NW, Suite 400
Washington, DC 20005
telephone: 202-296-1585
fax: 202-463-0474
e-mail: <jeanne@acpa.org>
URL: <http://www.acpa.org>

The ACPA is a nonprofit trade organization of U.S. manufacturers, formulators, and distributors of agricultural crop-protection and pest-control products. Since its creation in 1933, the ACPA has had four primary goals: (1) promoting sound legislation and science-based regulation for the crop protection industry; (2) raising the level of public understanding about the value of crop protection chemicals and other pesticides; (3) encouraging the continued use of sound scientific principles in research, development, and regulatory processes; and (4) addressing global issues affecting the export and safe use of U.S. pesticide products in other nations.

American Gas Association (AGA)
1515 Wilson Boulevard
Arlington, VA 22209
telephone: 703-841-8400
fax: 703-841-8406
e-mail: <info@aga.com>
URL: <http://www.aga.com>

The AGA was founded in 1918 for the purpose of serving as an advocate for natural gas local distribution companies (LDCs). The association now consists of nearly 300 companies involved in the gathering, marketing, distribution, and transmission of natural gas in North America. These companies account for more than 90% of all natural gas delivered in the United States.

Other objectives of the association include finding ways of increasing the demand for natural gas by promoting new natural gas products and technologies; coordinating the placement of trade and cooperative advertising; representing the natural gas industry to the general public; collecting, analyzing, and disseminating information on the natural gas industry; and conducting programs designed to improve the safety and efficiency of natural gas delivery.

American Petroleum Institute (API)
1220 L Street, NW
Washington, DC, 20005-4070
telephone: 202-682-8000
e-mail: <api@api.org>
URL: <http://www.api.org>

The API is the primary trade organization representing the U.S. petroleum industry. It was founded in 1919 and now consists of more than 300 companies involved in exploration, production, transportation, refining, and marketing of petroleum. The API has offices in Washington, DC, New York City, and 33 state capitals east of the Rocky Mountains. In the Southwest, Rocky Mountain area, West Coast, and Alaska, the API works in conjunction with regional and state oil and gas associations.

The institute's mission statement consists of four major points:
- meeting the nation's energy needs, developing energy sources, and supplying high-quality products and services;
- enhancing the environmental, health, and safety performance of the petroleum industry;
- conducting research to advance petroleum technology, and developing industry equipment and performance standards;
- advocating government decision-making to encourage efficient and economic oil and natural gas development, refining, transportation and use; promoting public understanding of the industry's value to society; and serving as a forum on issues affecting the petroleum industry.

Chlorine Chemistry Council (CCC)
1300 Wilson Boulevard
Arlington, VA 22209
telephone: 703-741-5000
                  800-CL-INFO-1
fax: 703-741-6084
e-mail: <Janet_Flynn@mail.CMAHQ.COM>
URL: <http://c3.org>

The CCC is a division of the Chemical Manufacturers Association. It was established in 1993 to promote the continuing and responsible uses of chlorine and chlorine-based products. Specific activities of the CCC involve the promotion of scientific research and risk assessment in public policies involving the use of chlorine, development and implementation of programs that "assure the practice and understanding of responsible stewardship for chlorine and chlorine-based products," sponsorship of research on the potential health and environmental effects of chlorine and chlorine-based products, creating better understanding of the health and safety benefits of chlorine in consumer products, and serving as a reliable source of information on chlorine and chlorine-based compounds.

Among the services provided by the CCC are the *Executive Newsletter*, a weekly digest of news about the council; a series of seven informational brochures on the role of chlorine in water disinfection, medicine, cleaning, and everyday life; an Internet homepage that provides information on current news events, public policy debates, scientific aspects of chlorine, and other technical information; and product stewardship publications, designed to aid customers in their stewardship efforts.

Color Pigments Manufacturers Association, Inc. (CPMA)
300 North Washington Street, Suite 102
Alexandria, VA 22314
telephone: 703-684-4044
fax: 703-684-1795

The mission statement of the CPMA contains four major goals:
- to serve the present and future needs of its member companies so they can better serve their customers, employees, and society;
- to promote the interests of the color pigments industry in North America;
- to responsibly influence member companies, government, and public actions for the benefit of all;
- to encourage high industry standards of environmental management.

Membership in the organization consists of business enterprises engaged in the manufacture and/or sale of pigment colors or intermediary chemicals involved in the production of such colors. Nearly a dozen CPMA committees are involved in safety and environmental issues surrounding the use of specific coloring chemicals, such as iron oxide, compounds of chromium, pththalocyanine pigments, and carbozole violet.

Composites Institute
1801 K Street, NW, Suite 600K
Washington, DC 20006-1301
telephone: 202-974-5200
fax: 202-296-7005
e-mail: <feedback@socplas.org>
URL: <http://www.socplas.org/BusinessUnits>

The Composites Institute is the largest division within the Society of the Plastics Industry. It was formed in 1945 to increase markets for composite materials. Members of the institute include resin, glass, and equipment suppliers; fabricators; and consultants and academics interested in the subject of composites.

Cosmetic, Toiletry, and Fragrance Association (CTFA)
1101 17th Street, NW, Suite 300
Washington, DC 20036
telephone: 202-331-1770
fax: 202-331-1969
e-mail: <ctfa@ctfa.org>
URL: <http://www.ctfa.org>

The CTFA was founded in 1894 for the purpose of protecting the freedom of members of the cosmetic, toiletry, and fragrance industries to compete in a fair and reasonable marketplace. The primary means of achieving this goal is through the promotion of voluntary industry self-regulation along with reasonable governmental requirements that ensure uniform national and international standards for the industry. The association currently consists of more than 500 member companies from throughout the world.

The Fertilizer Institute (TFI)
501 Second Street, NE
Washington, DC 20002
telephone: 202-675-8250
fax: 202-544-8123
e-mail: <information@tfi.org>
URL: <http://www.tfi.org>

TFI grew out of the National Fertilizer Association, founded in 1883. The association was created to increase industry business and to protect the fertilizer industry from unfair shipping monopolies, growing restraints on interstate business, and a confusing array of state fertilizer laws. A decade later, manufacturers from the Midwest formed the Fertilizer Manufacturers of the West and, in 1897, the Chemical Fertilizer Manufacturers Association was formed by companies in eastern states.

Today, more than 250 companies involved in plant food production, fertilizer distribution, secondary and micronutrient manufacture, equipment supply and engineering construction, brokering and trading, rail and barge transportation, and many other aspects of agriculture are members of TFI.

TFI's activities include working with the U.S. Congress to represent the interest of the fertilizer business; collecting, tabulating, and distributing statistics on fertilizer production and use; and promoting public outreach programs to increase general understanding of the fertilizer industry and the issues it faces.

TFI sponsors three major conferences each year and publishes a number of magazines, books, journals, and other materials. Included are a bimonthly magazine, *Dealer PROGRESS*, devoted to business management information and legislative and regulatory news; a monthly newsletter, *ACTION*, reporting on legislative, regulatory, and public affairs issues; the *TFI/NACA Agri-Dealer Action Fax Network*, providing information on legislative issues affecting retail dealers; the *Washington Wire*, a bulletin concerning regulatory and legislative news; and annual and semiannual reports, such as *Fertilizer Financial Facts*, *Fertilizer Production Cost Surveys*, and *Production Survey*.

National Paint and Coatings Association (NPCA)
1500 Rhode Island Avenue, NW
Washington, DC 20005
telephone: 202-462-6272
fax: 202-462-8549
e-mail: <npca@paint.org>
URL: <http://www.paint.org>

The NPCA is a voluntary, nonprofit trade association consisting of about 400 paint and coatings manufacturers, raw material suppliers, and distributors. Special activities of the organization are handled through a number of divisions and programs, including the Outreach Program of the Public Affairs Division; the Health, Safety and Environmental Affairs Division; the Government Affairs Division; and the Administration Division.

National Pest Control Association (NPCA)
8100 Oak Street
Dunn Loring, VA 22027
telephone: 703-573-8330
fax: 703-573-4116
e-mail: <Lederer@pestworld.org>
URL: <http://www.pestworld.org>

The NPCA was founded in 1933 and currently has more than 3,000 corporate members in the United States and other nations around the world. The organization's mission is "to communicate the role of our industry as protectors of food, health, property, and the environment, and [to] affect the success of our members through education and advocacy."

Among the activities of the association are contributions to the development and implementation of federal and state legislative and regulatory programs for pest control, the creation of verifiable training programs, improving technologies used by industries, and public and media relations.

Polyurethane Foam Association (PFA)
P. O. Box 1459
Wayne, NJ 07474-1459
telephone: 973-633-9044
fax: 973-628-8986
e-mail: <rluedeka@pfa.org>
URL: <http://www.pfa.org>

The mission of the Polyurethane Foam Association is to educate the general public about flexible polyurethane foam and to promote its use in manufactured and industrial products. Included in this mission is the goal of disseminating information on environmental, health, and safety issues related to polyurethane foam to the public, regulatory officials, business leaders, and the media. The PFA also provides members and customers with technical information on properties and performance of polyurethane foam in consumer and industrial products. The organization was founded in 1980.

Rubber Manufacturers Association (RMA)
1400 K Street
Washington, DC 20005
telephone: 202-682-4841
          800-220-7622
fax: 202-682-4854
e-mail: <kristen@tmn.com>
URL: <http://www.rma.org>

The RMA grew out of an organization called the New England Rubber Club, founded in Boston in 1900. By 1909, the club had been reorganized as the Rubber Club of America (RCA), and its offices moved to New York City. In 1929, the RCA was reorganized once more and named the Rubber Manufacturers Association. Today, the organization's members are jointly responsible for the production of approximately $20 billion in tires, tubes, roofing, sporting goods, and mechanical and industrial products. The organization is now divided into two major divisions, the General Products Group and the Tire Group. The RMA claims to be "the leading business advocate, resource and information clearinghouse for and about today's diversified rubber industry" (<http://www.rma.org/about.html>).

The Soap and Detergent Association (SDA)
475 Park Avenue South
New York, NY 10016
telephone: 212-725-1262
fax: 212-213-0685
e-mail: <webmaster@webcorp.com>
URL: <http://www.sdahq.org>

The SDA was founded in 1926 to advance public understanding of the safety and benefits of cleaning products and protecting the ability of its members to formulate products that best meet consumer needs. Today, the organization consists of about 135 manufacturers of household, industrial, and institutional cleaning products; the ingredients used in making those products; and the packaging in which those products are sold. SDA members are responsible for more than 90% of the cleaning products sold in the United States.

The SDA provides information about cleaning products and their environmental effects to its member companies, the general public, policymakers, childcare and health professionals, educators, and members of the media. A special mission of the organization is to provide information about the human and environmental safety of cleaning products and their ingredients; the safe and effective disposal of cleaning products and their packaging; and the contributions of cleaning products to personal and public health. This information is made available by way of publications, educational programs, reports, seminars, and conferences.

The Society of the Plastics Industry, Inc. (SPI)
1801 K Street, NW, Suite 600K
Washington, DC 20006
telephone: 202-974-5200
fax: 202-296-7005
e-mail: <feedback@socplas.org>
URL: <http://www.socplas.org>

The SPI consists of more than 2,000 member companies representing all segments of the plastics industry in the United States. These members include suppliers of raw materials, moldmakers, plastic processors, and groups and individuals involved in other aspects of the production of plastics. The organization was established in 1937 and has as its mission the promotion of "the continued development of the plastics industry" and the enhancement of "public understanding of plastics' contributions while meeting the needs of society." The society maintains regional offices in Boston, Chicago, Los Angeles, and Greenville, South Carolina. It also has 36 business units and service committees created to serve specialized needs of the plastics industry.

The Sulphur Institute (TSI)
1140 Connecticut Avenue, NW, Suite 612
Washington, DC 20036
telephone: 202-331-9660
fax: 202-293-2940
e-mail: <sulphur@access.digex.net>

The Sulphur Institute is an international organization representing the sulphur industry. The institute's purpose is to promote and expand the use of sulphur in all forms worldwide. TSI includes 13 members and 15 associate members, from the United States, Canada, and 9 other nations. TSI activities include participation in a variety of symposia, such as the International Symposium of Sulphur in Soils and

Sulphur Market Trends, Developments, and Opportunities; sponsorship of the annual Sulphur Phosphate Symposium; participation in international, national, and regional working groups on transportation and regulatory activities; and publication of an annual market report, *Sulphur Outlook,* and their monthly *President's Newsletter.*

Synthetic Organic Chemical Manufacturers Association (SOCMA)
P.O. Box 79160
Baltimore, MD 21279-0106
telephone: 202-414-4100
fax: 202-289-8584
e-mail: <hannan@socma.com>
URL: <http://www.socma.com>

The SOCMA is an association of more than 280 member companies involved in the production of 95% of the 50,000 chemicals produced in the United States. The typical member company is a small business, employing fewer than 50 employees and producing less than $50 million in annual sales. The SOCMA's goal is to develop products and services that will enable member companies to "improve their performance, effectively manage environmental, health, safety and trade regulations, successfully voice concerns to the government, as well as create opportunities to profit commercially."

An important part of their work is carried out through 21 committees and subcommittees, such as those involved with hazardous wastes, international trade, Responsible Care®, occupational safety and health, and the Toxic Substances Control Act.

CHAPTER TEN
Print and Electronic Resources

T his chapter lists resources that provide up-to-date information on current developments in the chemical sciences. Those resources are largely of two kinds: journals and electronic publications. These publications are published more frequently than are books and can, therefore, provide the most recent information available on most topics.

The chapter is divided into two sections: Print Resources and Electronic Resources. In many instances, a magazine or journal that appears in print also has an electronic counterpart. In other cases, either a journal or an electronic Internet site exists independently of a counterpart in the other format.

PRINT RESOURCES

American Scientist
P.O. Box 13975
Research Triangle Park, NC 27709
telephone: 919-549-0097
        800-282-0444
fax: 919-549-0090
e-mail: <subs@amsci.org>
URL: <http://www.amsci.org/amsci/subscription/subscription.html>

Subscription rates: Rates depend on geographic location and subscription type. An individual subscription in the United States is $28 for one year, $50 for two years, or $70 for three years. An institutional subscription in the United States is $50 per year. Rates are slightly higher outside the United States.

American Scientist is a publication of Sigma Xi, the Scientific Research Society. It is published bimonthly and features a variety of articles about science and technology. Many of the articles are written by scientists and engineers working at the forefronts of their fields, thus providing valuable overviews of the direction and pace of progress in these fields. An interesting feature of the journal is its annual review of books, software, and other products written by children for younger readers.

   American Scientist has also published two books of readings collected from previously published articles in the journal. They are *Exploring Evolutionary Biology* and *Exploring Animal Behavior.*

The Chemical Educator
Springer-Verlag New York
P.O. Box 2485
Secaucus, NJ 07096-2485
telephone: 800-SPRINGER (777-4643), extension 526
e-mail: <chedr@springer-ny.com>
URL: <http://journals.springer-ny.com/>
Subscription rates: $24.95 per year

The Chemical Educator, which first appeared in 1996, is designed for chemical education professionals. Its major sections deal with topics such as coursework and activities in the chemistry classroom; laboratories and demonstrations; computers in chemistry; chemistry and history, issues involving chemical education in high school today; plus topics discussed in a chemical education forum. Journal articles can also be downloaded from the Internet using the Adobe Acrobat Reader. Six issues are published each year.

Chemical Week
888 Seventh Avenue, 26th Floor
New York, NY 10106
telephone: 212-621-4929
fax: 212-621-4946
e-mail: <circ@chemweek.com>
URL: <http://www.chemweek.com>
Subscription rates: Rates depend on geographic location. A subscription in the United States is $115 per year.

Chemical Week is a wide-ranging magazine dealing with virtually every aspect of the chemical industry, from reports on research to company profiles. The five major sections of the magazine are Specialty Chemicals; Markets and Economics; Technology and Environment; Companies, Management, and People; and Services to the Industry.

   In addition to the magazine itself, *Chemical Week* publishes a large selection of books dealing with the chemical industry; newsletters; buyers' guides and service directories; *Chemical Week* reprints; and CD-ROMs and binders containing full transcripts of conferences sponsored by *Chemical Week*.

Chemistry & Industry
Society of Chemical Industry
15 Belgrave Square
London SW1X 8PS
U.K.
telephone: +44 (0) 171-235-3681
fax: +44 (0) 171-823-1698
e-mail: <members@chemind.demon.co.uk>
URL: <http://sci.mond.org>
Subscription rates: Rates depend on geographic location. A subscription in the United States is $422 per year.

Chemistry & Industry is published twice a month and carries news about the chemical industry around the world as well as reports on recent developments in chemical research. Each issue focuses on two or three special areas of interest, such as plant biotechnology, biopolymer production, molecular approaches to improving wheat quality, health impacts of indoor air pollution, clean air issues, microbial biofilms, and so on. Although some portions of the journal are accessible only to professional scientists, the general articles are understandable to the educated layperson. The magazine is well written, beautifully illustrated, and a superb way to stay in touch with the most recent progress in chemical research.

   Chemistry & Industry also has a very useful Internet site that provides a review of the current issue, a section called "Today's News in Chemistry," archival information, a job-search section, and a collection of news and feature articles from the world of chemistry, called "Only Connect."

Chem Matters
American Chemical Society
P.O. Box 2537
Kearneysville, WV 25430
telephone: 800-209-0423
fax: 800-209-0064
URL: <http://www.acs.org/edugen2/education/acapgms/edupgms/catalog/orders/cm_ord.htm>
Subscription rates: 1–4 copies of the student magazine (product #HS29), mailed to the same address, are $8.00 each; 5 or more copies are $3.50 each. A separate teacher's guide (product #HS28) is available for $3.00 per subscription.

Chem Matters is designed to show high school students the role of chemistry in everyday life. It is edited by a high school teacher and includes feature stories, puzzles, and real-life mystery tales that describe the work that chemists do. The supplementary classroom guide provides additional information on magazine articles, suggested laboratory experiments and classroom demonstrations, and questions for students.

   All back issues of *Chem Matters,* from its first publication in 1983 through the end of 1995, are also available on a CD-ROM that contains a searchable index, based on either a topic or a keyword. The CD-ROM costs $54.95 for the individual-user version (product #HS35A) and $84.95 for the Network/Library version (product #HS35B).

CHEMTECH

American Chemical Society
1155 16th Street, NW
Washington, DC 20036
telephone: 614-447-3776
                800-333-9511
fax: 614-447-3671
e-mail: <chemtech@acs.org>
URL: <http://pubs.acs.org/journals/chtedd/about.html>
Subscription rates: Rates depend on geographic location, membership status, and subscription type. In the United States, an individual member subscription is $46 per year; an individual subscription is $93; an institutional nonmember subscription is $450; and a student member subscription is $23.

CHEMTECH covers a whole range of topics of interest to those involved in chemistry-related industries, including resources needed to initiate innovation in chemical technology, research and development on new technologies, and marketplace issues involved in the commercial production of such technologies. The magazine contains both feature articles and regular departments focusing on chemical-related technologies; ideas for the commercial development of technologies; and environmental, research, and legal concerns.

Chemunity News

American Chemical Society
Education Division
1155 16th Street, NW
Washington, DC 20036
telephone: 800-ACS-5558 (press "0" and ask for Shirley Mundle in the Education Division)
URL: <http://www.acs.org/edugen2/education/acapgms/edupgms/chemnew/chemnew.htm>
Subscription rates: The newsletter is free upon request.

Chemunity News is published by the ACS Education Division and is designed for teachers, industrial chemists, members of the ACS, and anyone interested in the activities of the ACS Education Division.

Discover

Circulation Department
114 Fifth Avenue
New York, NY 10011-5690
telephone: 515-247-7569
                800-829-9132
e-mail: <discover@enews.com>

URL: <http://www.enews.com/magazines/discover/>
Subscription rates: $24.95 per year in the United States; U.S. $35.95 plus 7% GST in Canada.

Walt Disney Corporation's *Discover* is a general-interest magazine that fills the needs of science enthusiasts much like *Sports Illustrated* does for sports fans. *Discover* is published 12 times a year and carries stories that attempt to make advances in science and technology understandable to the general public. The magazine claims to be "the only magazine that covers the entire world of science."

In addition to the magazine itself, *Discover* provides a number of other print and nonprint resources. For example, it maintains a *Discover Magazine Virtual Bookstore* from which books can be purchased online, and it has produced two video series, *Great Minds of Science* and *Secrets of Science*. *Discover* also appears as a monthly television program that focuses on some specific topic in science and technology. In one month, for example, topics of the television show were "Air Crashes," "Mutations," "Size and Scale," and "Natural Born Killers."

Education in Chemistry
Turpin Distribution Services Limited
Blackhorse Road
Letchworth, Herts SG6 1HN
U.K.
telephone: +44 (0) 146-267-2555
fax: +44 (0) 146-248-0947
e-mail: <turpin@rsc.org>
URL: <http://chemistry.rsc.org/rsc/ed_chem.htm>
Subscription rates: U.S. $210 or £116.00 per year

Education in Chemistry is published by the Royal Society of Chemistry for teachers of chemistry at all levels, from presecondary to college. Its seven major departments include an editorial column; a letters-to-the-editor column; feature articles dealing with topics of general chemical interest; "Distillates," which reports on recent chemical research; reviews of books, computer software, and videos relating to chemical education; "Endpoint," a space for individual observations and comments; and "InfoChem," a pupil supplement designed for 16- to 18-year-old students studying chemistry.

in Chemistry
American Chemical Society
1155 16th Street, NW
Washington, DC 20036
telephone: 800-227-5588
e-mail: <SAprogram@acs.org>
Subscription rates: *in Chemistry* is provided as a benefit to members in ACS student affiliate groups.

in Chemistry is the official magazine published for student affiliates of the American Chemical Society. It contains articles dealing with special topics in chemistry, career fields, professional development and workplace issues, guest editorials and interviews, student-affiliate chapter activities, and other topics of special interest to students interested in chemistry.

Journal of the American Chemical Society
American Chemical Society
1155 16th Street, NW
Washington, DC 20036
telephone: 614-447-3776
             800-333-9511
e-mail: <jacs@mail.utexas.edu>
URL: <http://pubs.acs.org/journals/jacsat/about.html>
Subscription rates: Rates depend on geographic location, membership status, and type of subscription. In the United States, an individual member subscription is $125 per year; a student member subscription is $93.75; and an individual nonmember subscription is $1,695. The journal is also published and updated bimonthly on disk. The rate for a member with a hard copy subscription is $68 per year and without a hard copy subscription is $151. The rate for nonmembers with a hard copy subscription is $725 per year and without a hard copy subscription is $1,966.

The *Journal of the American Chemical Society* was founded in 1879 and is arguably the most prestigious general journal on chemistry in the world today. Each year, more than 13,500 pages of research, reports, book and software reviews, correspondence, and comment are published.

Journal of Chemical Education
Subscription and Book Order Department
P.O. Box 606
Vineland, NJ 08360
telephone: 800-691-9846
fax: 609-696-2130
e-mail: <jchemed@aol.com>
URL: <http://jchemed.chem.wisc.edu/Journal/index.html>
Subscription rates: An individual member subscription in the United States is $35 per year. Non-U.S. subscription is $45 per year.

Every month, for over three-quarters of a century, the *Journal of Chemical Education* *(JCE)* has been publishing news and articles of interest to chemical educators. The journal contains seven major sections: Chemical Education Today; In the Classroom: Writing; In the Classroom: Experiments and Demonstrations; In the Laboratory; Information, Textbooks, Media, and Resources; Chemistry Everyday for Everyone; and Research: Science and Education. The journal is primarily designed for high school and college teachers of chemistry but has some articles of interest to students.

   In addition to the journal, the *JCE* publishes a number of books, such as 10- and 25-year cumulative indexes; *Modern Experiments for Introductory College Chemistry*; *Safety in the Chemical Laboratory*, volumes 2–4; *Handbook for Teaching Assis-*

tants; and *Chemistry of Art*. The *JCE* also publishes offprints such as "Solid State Chemistry," "Radiation Chemistry," "Electrochemistry," "Inorganic Photochemistry," and "Chemistry of the Food Cycle."

Kids Discover
P.O. Box 54205
Boulder, CO 80322-4205
telephone: 212-242-5133
Subscription rates: $19.95 per year. Single copies are available at $3.00 per copy. Telephone for multiple copies.

Kids Discover is published 10 times a year, with each issue focusing on a single topic. Some topics that have been covered in the past include oil, the solar system, the brain, garbage, endangered species, and light.

Nature
(In the United States)
Nature America, Inc.
345 Park Avenue South, 10th Floor
New York, NY 10010-1707
telephone: 800-524-0384
fax: 615-377-0525
e-mail: <subscriptions@nature.com> (for subscriptions only)
URL: <http://www.nature.com>
Subscription rates: Rates depend on geographic location and subscription type. In the United States, an individual subscription is $145 per year; an institutional subscription is $495; and a student subscription is $80.

Nature is the United Kingdom's counterpart to *Science* in the United States—a highly respected journal covering all areas of science. It provides news and research reports that are accessible to the general reader with a moderate background in science, along with more esoteric reports on research in all fields of science.

The journal now publishes specialized magazines in the fields of medicine (*Nature Medicine*), genetics (*Nature Genetics*), biology (*Nature Structural Biology*), and biotechnology (*Nature Biotechnology*).

The family of *Nature* journals is also available online. This URL is open to the general public and provides weekly tables of contents, summaries, news articles, access to *Nature* archives, and job information.

New Scientist
1150 185th Street, NW, #725
Washington, DC 10036
telephone: 888-800-8077
fax: 202-331-2082
e-mail: <newscidc@soho.ios.com>
URL: <http://www.newscientist.com/about/ns.html>
Subscription rates: In the United States, a subscription is $140 for one year; $252 for two years; $336 for three years; a student subscription is $70 per year.

New Scientist is published 51 times a year. It was begun in 1956 "for all those men and women who are interested in scientific discovery and in its industrial, commercial and social consequences." In its first two decades, the magazine was a powerful advocate for the analysis of the social, political, economic, and ethical implications of scientific research and technology. Over the last decade, it has tempered its social agenda to some extent, but it is arguably the best single source for a rounded view of the place of science and technology in modern society, as well as being an excellent source of information on developments in both areas.

Popular Science
P.O. Box 51282
Boulder, CO 80322-1282
telephone: 800-289-9399
URL: <http://www.popsci.com/support/about.html>
Subscription rates: $17.94 per year.

Popular Science has been published continuously since 1872. It is one of the premier magazines attempting to explain developments in science and, especially, technology to the average reader. Some of the areas covered by the magazine include automotive science and technology, consumer electronics, computers and software, home technology, photography, and aviation and space.

Reaction Times
American Chemical Society
1155 16th Street, NW
Washington, DC 20036
telephone: 614-447-3776
              800-333-9511
e-mail: <rxnt@acs.org>
URL: <http://www.acs.org/memgen/rxntimes/rxntimes.htm>

Reaction Times is published four times each school year. It is designed for college students in chemistry and chemical engineering and for others interested in the diversity of the chemical sciences and careers in the chemical sciences.

Science
American Association for the Advancement of Science
1200 New York Avenue, NW
Washington, DC 20005
telephone: 202-326-6400
e-mail: <webmaster@aaas.org>
URL: <http://www.aaas.org>
Subscription rates: Rates depend on geographic location, membership status, and subscription type. In the United States, a professional member subscription is $105 per year; and a subscription including access to *Science Online,* the electronic version of *Science*, is $117; student subscriptions are $58 and $70, respectively.

Science is one of the two premier scientific journals for general readers around the world (*Nature* being the other). The magazine contains seven major sections: News and Comment; Research News; Perspectives; Articles; Reports; Technical Comments; and Web Feature. In addition, its regular departments include "This Week in *Science*," a brief overview of the magazine; an editorial column; a letters-to-the-editor column; "Random Samples," which presents brief news notes; book reviews; and "Tech.Sight: Products," which provides information on new technology and products.

Science has an extensive Internet Web site that includes two major sections: Science NOW and Science's Next Wave. Science NOW includes "Current Stories" and "Week in Review," both of which present stories from the world of science. Science NOW also contains a search function that allows a limited search of the journal's archives and current issues. Next Wave includes "Going Public," an open forum for discussion of topics in science; "New Niches," a feature on alternative careers in science; "In the Loop," a news section; "Tooling Up," a column on science career advice; "Signposts," which provides links to other Web pages; and "Wavelengths," which provides an opportunity to interact with the editors of *Science*.

Science News
Science News Subscriptions
231 West Center Street
P.O. Box 1925
Marion, OH 43305
telephone: 202-785-0931 (collect)
              800-552-4412
e-mail: <scinews@sciserv.org>
URL: <http://www.sciencenews.org>
Subscription rates: $49.50 for one year or $88.00 for two years

Science News is a publication of Science Service, a nonprofit corporation based in Washington, DC. The publication started out in 1921 as a series of news releases dealing with scientific issues. It then evolved into the current magazine, which is published 51 times a year. Science Service also administers the annual Westinghouse Electric Corporation's Science Talent Search and the International Science and Engineering Fair.

The magazine is normally a 16-page publication that summarizes about a dozen important breakthroughs in all fields of science, along with brief Research Notes from two or three major fields of science and two articles covering some major topics in science. The magazine also includes a book review and letters-to-the-editor section. *Science News* is arguably the best single publication reporting on advances in science that is available to the general reader.

The Sciences
Publications
New York Academy of Sciences
2 East 63rd Street
New York, NY 10021
telephone: 212-838-0230, extension 342
              800-843-6927
fax: 212-888-2894
e-mail: <publications@nyas.org>

URL: <http://www.nyas.org/sciences.htm>
Subscription rates: Rates depend on geographic location. In the United States, a subscription is $21 for one year, $37 for two years, or $48 for three years. Rates are slightly higher outside the United States.

The Sciences describes itself as "the cultural magazine of science . . . combin[ing] the literary and aesthetic values of a fine consumer magazine with the authority of a scholarly journal to lead the reader in an enlightening exploration of the world of science." Indeed, the journal does provide a broad and complex overview of scientific research that is probably not presented as well in any other general-interest magazine. The appeal of the magazine is reflected in its having won National Magazine Awards in 1986, 1988, 1989, 1991, and 1996, along with a host of other publishing honors.

Scientific American
415 Madison Avenue
New York, NY 10017-1111
telephone: 515-247-7631
            800-333-1199
fax: 515-246-1020
e-mail: <customerservice@sciam.com>
URL: <http://www.sciam.com>
Subscription rates: A subscription in the United States is $34.97 per year. Subscriptions outside the U.S. are $49.00 per year.

Scientific American, which was first published in 1845, is the oldest and most prestigious scientific journal written for the general reader. This monthly magazine includes articles covering every field of science and technology, written at a level that an educated and motivated high school, college student, or adult can understand.

Scientific American now has an extensive Internet home page that includes the current issue, archives of past issues, weekly features, breaking news stories, interviews with scientists, question-and-answer sessions between readers and specialists in all fields of science, and links to other sites containing further information on magazine articles.

Technology Review
P.O. Box 489
Mount Morris, IL 61054
telephone: 800-877-5320
e-mail: <trsubscriptions@mit.edu>
URL: <http://web.mit.edu/technreview/www.customer.html>
Subscription rates: Rates depend on geographic location. A subscription in the United States is $32 per year; for a subscription in Canada add $6, and for other nations, add $12.

Technology Review is published eight times a year by the Association of Alumni and Alumnae of the Massachusetts Institute of Technology. The magazine carries popular accounts of important technological advances, often with special emphasis on social, political, economic, and ethical issues related to the technology. Typical topics of stories include complexity theory, microwave transmission of solar energy, the landmine controversy, and recycling of household garbage.

Some regular departments are a letters-to-the-editor column, the MIT Reporter ("An Inside Look at Research at MIT"), Trends, The Economic Perspective, and The Humane Engineer.

Today's Chemist at Work
American Chemical Society
Member and Subscriber Services
P. O. Box 3337
Columbus, OH 43210
telephone: 614-447-3776
800-333-9511
fax: 614-447-3671
e-mail: <service@acs.org>
URL: <http://pubs.acs.org/journals/tcwoe7/about.html>
Subscription rates: Rates depend on geographic location and membership status. In the United States, a member subscription $17 per year, and a nonmember subscription is $95. The journal is distributed free to chemists working in industry.

Today's Chemist at Work is published 11 times a year. It deals with personal, financial, and career issues of professional chemists. Its articles include topics such as new computer techniques, peripherals and software, instrumentation, testing methods, formulations, environmental regulations, and communication systems.

WonderScience
American Chemical Society
1155 16th Street, NW
Washington, DC 20036
telephone: 202-872-6366
                  800-209-0423
e-mail: <r_foster@acs.org>
URL: <http://www.acs.org/education/currmats/wondsci.html>
Subscription rates: Subscriptions are $12.00 per year for a single year. Reduced rates are available for more than one subscription at the same address.

WonderScience is published in cooperation with the American Institute of Physics. It is an attractive, four-color magazine containing creative activities that elementary school teachers can use in the classroom. It features a variety of hands-on activities that can be carried out with inexpensive, safe, and easily accessible materials found at home or in the grocery store. *WonderScience* is published eight times a year, with four issues in September and four in January. A complete set of back issues for the last 10 years of the magazine is also available.

ELECTRONIC RESOURCES

Electronic sources can be produced and modified much more quickly than can print sources. As a result, they tend to appear, change locations, and disappear much more quickly than is the case with print journals and magazines. Some of the sources listed below may no longer exist in the same format that they did when this book was written, and they may have been replaced by other sources. The reader is advised to make use of standard search methods to locate sources of chemical information on the Internet and not to rely entirely on those listed below.

The Armchair Scientist

e-mail: <L.Crudeli@areacom.it>
URL: <http://www.areacom.it.html.arte_cultura/loris/armchair.html>

An electronic journal based in Italy, *The Armchair Scientist* is an interesting effort to present news about developments in science and technology, much of which is provided by its readers. It attempts to present this information in "an attractive and unimposing way" while, at the same time, trying to "stimulate the quest for a new way to do science." The four major sections of the journal are The Science Show, The Technology Show, Heavy Science, and New Theories. Some of the regular departments include a forum for mail and announcements; "Sharing Knowledge," which includes information and commentary from readers; "The Treasure Chest," an index of previous issues of the journal; "Events," a list of major science and technology events; and "Web Links," a list of useful Web sites. The Web site is maintained by Loris Crudeli.

ChemDex

e-mail: <webelements@Sheffield.ac.uk>
URL: <http://www.shef.ac.uk/chemistry/chemdex/>

ChemDex is maintained by Mark Winter at the University of Sheffield, England. The Web site is a superb source of chemical information and links to other important chemistry Web sites around the world. Sheffield's periodic table section is an essential beginning point for anyone wanting to find data on the chemical elements.

Chemical Engineering

URL: <http://www.che.ufl.edu/WWW-CHE/index.html>
e-mail: <www-che@www.che.ufl.edu>

This Web site carries extensive information about meetings and conferences in chemical engineering. Some of the subtopics listed include biomedical, biotechnology, ceramics, chemical producers and suppliers, databases, education resources, electrochemical, environment, fluid mechanics, forest products, heat transfer, nuclear, particle technology, petrochemicals and fuels, polymers, process control, process design, and standards. An additional section provides links to academic institutions, professional organizations, research organizations and laboratories, and consultant and service providers. Another section provides access to a variety of relevant information sources, such as catalogs, newsgroups, bulletin boards, periodicals, and patents and patent information. The Web site is maintained by Dale Kirmse, Process Improvement Laboratory, Chemical Engineering Department, University of Florida.

Chemical On-line Presentations, Talks and Workshops
e-mail: <rzepa@ic.ac.uk>
URL: <http://www.ch.ic.ac.uk/talks/>

This site makes available to the general public the opportunity to post any talk, poster, or workshop presentation that has been made in the past or that will be made in the future. Examples of such presentations currently on the Web site include "Chemistry and the Web: Hyperactive Molecules," "Hyperactive Chemistry," "Non-Classical Hydrogen Bonding in Chiral Recognition and Mechanism," "Ionization Threshold Shifts in Covalently Linked Dimers," "The World-Wide Web as a Scientific Enabler," "Lab Safety for Students," "HIV-1: What Is It and How Does It Cause AIDS?" and "Non-Classical Hydrogen Bonds Illustrated Using the World-Wide Web." The Web site is maintained by Henry Rzepa, Imperial College, London.

Chemistry Information on the Internet
URL: <http://hackberry.chem.niu.edu/cheminf.htm>
e-mail: <admin@hackberry.chem.niu.edu>

Chemistry Information on the Internet is one of the most nicely designed, well-maintained chemistry sites on the Internet. It is maintained by Steven Bachrach at Northern Illinois University. The site is divided into the following categories: New and Exciting Chemistry Web Developments, Chemistry Databases and Related Information, Electronic Journals, Chemistry World-Wide Web Sites, Chemistry Gopher Sites, and Other Lists of Chemistry Resources on the Internet. This site is a good one from which to begin a more detailed search for other sites and topics in the field of chemical research.

Informatics in Chemistry
URL: <http://macedonia.nrcps.ariadne-t.gr>

This site is located at the Institute of Physical Chemistry, a division of NCSR "Demokritos," in Athens, Greece. The site has been in existence since 1990 and has developed an extensive database of chemical information in more than 60 areas, including agricultural chemistry, aquatic chemistry, atmospheric chemistry, chemical education, chemistry in art, corrosion, forensic chemistry, fuel chemistry, marine chemistry, materials science, molecular modeling, polymer chemistry and technology, soil chemistry, stellar chemistry, supramolecular chemistry, surface chemistry, safety, and toxicology.

The Institute for Scientific Information
3501 Market Street
Philadelphia, PA 19104
telephone: 800-336-4474 (press "2" at prompt)
fax: 215-386-2911
e-mail: <webmaster@isinet.com>
URL: <http://www.isinet.com>
The Institute for Scientific Information (ISI) is arguably the most complete and comprehensive source of scientific information in the world. It maintains a database

covering more than 16,000 international journals, books, and proceedings in the sciences, social sciences, and arts and humanities. It indexes complete bibliographic data, cited references, and author abstracts for every item in its database. This database serves as the foundation for a number of specialized products, such as customized alerting services, chemical information products, and retrospective citation indexes. These resources are available in a variety of formats, including print, diskette, CD-ROM, magnetic tape, online access, and the Internet. The ISI describes its ultimate objective as follows: " . . . to provide immediate desktop access to the most significant scientific literature. In its entirety. To anyone, anytime, anywhere."

Internet Chemistry Resources: Guide to Internet Resources
URL: <http://www.chem.rpi.edu/icr/chemres/chemres_08.html>
e-mail: <wardej@rpi.edu>

This Web site is maintained by Rensselaer Polytechnic Institute and provides links to a number of other chemistry-related Web sites. This page provides hyperlinks to catalogs and publishers, databases and data collections, e-mail servers, ftp resources, gophers, online search services, periodicals and conference proceedings, software, teaching resources, and miscellaneous Web resources. The site also has a search function.

Internet Resources for Science and Mathematics Education
URL: <http://www.inform.umd.edu/UMS+State/UMD-P...s/MCTP/Technology/MCTP_WWW_Bookmarks.html>
e-mail: <to2@umail.umd.edu>

This site provides one of the most extensive collections of resources of potential interest to teachers of science and math at all levels of education. It is a service of the Maryland Collaborative for Teacher Preparation and is maintained by Professor T. C. O'Haver of the Department of Chemistry and Biochemistry at the University of Maryland at College Park.
   This site is divided into the following categories: New and Notable, Education, Mathematics, Integrated Science and Mathematics, Chemistry, Computers and Technology, Downloadable Software, Multicultural Connections, History, and Art and Music.

The Learning Matter of Chemistry: Chemistry Information on the Internet
URL: <http://yip5.chem.wfu.edu/yip/cheminfo/cheminfo.html>
e-mail: <ylwong@aol.com>

This site is maintained by Dr. Yue-Ling Wong and contains a huge database of Web servers. A search function is available, along with extensive lists of gopher servers, ftp servers, telnet servers, news servers, and other sources of information about chemistry on the Internet.

naturalSCIENCE
URL: <http://naturalscience.com/ns/nshome.html#Top>
e-mail: <publisher@naturalscience.com>
naturalSCIENCE claims to be "an independent [electronic] journal published without government subsidy or lobby group support." It provides updates on developments in

science and technology, along with social issues relating to those developments. A typical issue includes one feature story and the following departments: "What Else Is New?" Previous Cover Stories; Editorials; Commentaries; Articles; Research Notes; Reviews; Letters to the Editor; From a Correspondent; Book Review; News Capsules; Call for Papers; Help Wanted; Science-Related Web Sites; Internet Coverage; Internet Search; Quotations; Author Guide; and Submit News Item.

Only Connect
See *Chemistry & Industry* in the Print Resources section of this chapter.
e-mail: <webmaster@chemind.demon.co.uk>
URL: <http://ci.mond.org/9702/strands.html>

Science NOW
See *Science* in the Print Resources section of this chapter.
e-mail: <science-feedback@forsythe.stanford.edu>
URL: <http://www.sciencenow.org/>

Science's Next Wave
See *Science* in the Print Resources section of this chapter.
e-mail: <nextwave-feedback@forsythe.stanford.edu>
URL: <http://www.nextwave.org/>

What's New in the World?
URL: <http://www.exploratorium.edu/learning_studio/news/>
e-mail: <jims@exploratorium.edu>

This site is an electronic product of San Francisco's famous hands-on science museum, the Exploratorium. The site includes both news and feature stories as well as interactive exhibits.

The World-Wide Web Virtual Library: Chemistry
URL: <http://www.chem.ucla.edu/chempointers.html>
e-mail: <mik@chem.ucla.edu>

This highly respected Web site is divided into two general categories: WWW Chemistry Sites and Other Chemistry Services. The former category includes academic institutions, nonprofit organizations, commercial organizations, other lists of chemistry resources and www virtual libraries special projects, chemistry software ads, specialty ads, and chemical industry consultants. The latter category includes chemistry gopher services, chemistry ftp servers, and chemistry and biochemistry usenet news groups. The Web site is maintained by Max Kopelvich, Department of Chemistry, University of California at Los Angeles.

YAHOO!
URL: <http://www.yahoo.com/yahoo/Science/Chemistry>

YAHOO! is one of the major search engines on the Internet. It maintains a specific location through which an individual can search for subjects or authors of books in the field of chemistry. It also lists a number of directories and indexes related to chemistry. A few of the distinct topics under which one can search are academic papers, alchemy, biochemistry, chemical and biological weapons, chemical engineering, chemometrics, chromatography, clinical chemistry, electrochemistry, elements, journals, molecular databases, organic chemistry, organizations, and scientific consultants.

CHAPTER ELEVEN
Glossary

This glossary does not contain the majority of chemical terms learned by students in the first year of the standard high school or introductory college chemistry course. It does include terms less likely to have been introduced in such courses as well as more advanced terms with which the student with one year of chemistry may not be familiar. In cases where the familiarity of a term is in doubt, it has been included in the following list. For a more detailed discussion of any term about which the reader may be uncertain, please see any introductory chemistry textbook or any standard reference, such as *Hawley's Condensed Chemical Dictionary, 12th edition* (revised by Richard J. Lewis, Sr.; Van Nostrand Reinhold Company).

abiotic   Not alive or not derived from a living organism.

acetylsalicylic acid   A compound found naturally in the bark of the willow tree and from which aspirin is made, usually by the replacement of one acidic hydrogen by a sodium atom, resulting in the production of sodium acetylsalicylate.

acoustic   Having to do with sound or the sense of hearing.

active block   One of two large domains in the molecular structure of a superconductor in which superconductivity is thought to be generated.

adenosine diphosphate (ADP)   A molecule consisting of one adenine unit, one ribose unit, and two phosphate units. The molecule is an intermediary product formed in cells during the production of adenosine triphosphate (ATP), the molecule that supplies energy for many biochemical reactions.

adenosine monophosphate (AMP)   A molecule consisting of one adenine unit, one ribose unit, and one phosphate unit. The molecule is a starting material in cells during the production of adenosine triphosphate (ATP), the molecule that supplies energy for many biochemical reactions.

adenosine triphosphate (ATP)   A molecule consisting of one adenine unit, one ribose unit, and three phosphate units. The molecule stores large amounts of energy in its bonds, energy that it transfers to other compounds to make possible a variety of chemical changes within living organisms. ATP is generated in cells when two phosphate groups are added successively to a single molecule of adenosine monophosphate (AMP).

alpha helix   One of the two common three-dimensional forms taken by a polypeptide chain; it has the general appearance of a spiral staircase or a "slinky" toy.

ambient conditions   Environmental or surrounding conditions.

amino acid residue   The portion of an amino acid that remains after that acid has combined with other amino acids to form a peptide, oligopeptide, polypeptide, or similar structure.

amphiphilic   In general, to have the ability to attach to both polar (water-soluble) and nonpolar (water-insoluble) surfaces. For example, an amphiphilic molecule is one with a hydrophilic (water-attracting) and hydrophobic (water-repelling) end.

analog   In general, anything that is similar to something else in most, but not all, respects. In chemistry, an analog (also analogue) is a compound that is similar in many ways to some other compound, but that differs from it in some specific way, such as the presence or absence of one atom or group of atoms.

antigen   Any substance capable of stimulating an antibody response.

antiserum   A serum that contains antibodies.

arachidonic acid   An organic acid with the systematic name of 5,8,11,14-eicosatetraenoic acid and the formula of $CH_3(CH_2)_4(CH=CHCH_2)_4(CH_2)_2COOH$. The compound is found naturally in the liver, brain, and glands, and is the precursor of prostaglandins.

aromatic   In chemistry, a compound or structure with a benzene-type ring, consisting of six carbon atoms joined to each other in a resonance structure with three double bonds.

aromatization   A chemical reaction in which a straight chain is converted into an aromatic-like ring compound or structure.

ATP   See *adenosine triphosphate.*

bar   A unit of pressure equal to 10^5 pascal.

beta sheet   One of the two common three-dimensional structures taken by a polypeptide chain. It has a wavy, accordion-shaped appearance. It is also called a pleated sheet.

biosensor Any device that is able to detect, record, and transmit information about a changeable characteristic of a living organism (such as its temperature).

bipolar particle A particle in which electrons are unequally distributed such that one portion is more positive (or more negative) than another portion.

blood-clotting cascade (BCC) The series of chemical reactions that result in the clotting (or coagulation) of blood.

blood-clotting factor A term used to describe any of the substances involved in the complex series of reactions by which blood clots. It includes enzymes, proenzymes, vitamin K, and the calcium ion (Ca^{2+}).

bohrium (Bh) The name suggested by the International Union of Pure and Applied Chemistry for element number 107.

"bubble" system In systems for the control of pollution, any agreement whereby the limits established are applied to the sum of all emissions by all individual bodies within some given region, or "bubble."

buckminsterfullerene An allotropic form of carbon discovered in 1985 that consists of 60 carbon atoms arranged in a soccer-ball-like structure consisting of hexagons and pentagons.

buckyball A nickname for a buckminsterfullerene molecule.

cavitation The formation of bubbles in a liquid when that liquid is exposed to some kind of stress, such as passing a sound wave through the liquid.

chalcogen A generic name for the elements that make up Group 16 in the periodic table: oxygen, sulfur, selenium, tellurium, and polonium.

chalcogenide A binary compound containing a chalcogen.

chaperone protein A protein that becomes physically or chemically attached to some other species (such as a metal ion) and is responsible for transporting that species through some system, such as a living cell.

charge-reserve block A large domain within the molecular structure of a superconductor in which charges are thought to be produced before being transferred to the active block of the structure.

chelate A coordination compound in which a central metal ion is attached to two or more nonmetal atoms or groups of atoms, known as ligands.

chemical weapon Any chemical whose purpose is to cause death, injury, or temporary loss of performance to people.

chlorofluorocarbons (CFCs) A group of compounds in which all of the hydrogen atoms of an alkane have been replaced by one or more halogen atoms.

chromophore A group of atoms in an aromatic compound that is responsible for the color of that compound.

clathrate A type of compound in which atoms or molecules of one substance are completely enclosed within the molecules of a second substance.

cofactor Any substance associated with an enzyme to enable its function.

copolymer A polymer that is produced by the reaction between two or more different kinds of monomers.

cyclase An enzyme that catalyzes reactions by which a straight chain is converted into a ring structure.

cyclization The process by which a straight chain of carbon atoms is converted into a ring structure.

dalton One atomic mass unit (a unit of mass equal to 1/12 the mass of an atom of carbon-12).

dendrimer A polymer-like substance whose growth typically begins from some central core structure and builds outward in a manner similar to the way that a tree grows. Dendrimers exist in more than 50 different shapes, and differ from polymers in that their growth can be precisely controlled and directed.

diamond anvil cell A device (invented by Percy Bridgman in the early 1900s) for exposing small samples of matter to very high pressures. The material to be studied is placed between two flat plates made of diamond and exposed to pressures up to a million atmospheres or more.

dihydrogen bond A new type of bonding (announced by researchers in 1996) that exists between two hydrogen atoms, one carrying a slightly positive charge and one carrying a slightly negative charge.

dinitrogen A form of nitrogen in which each nitrogen molecule consists of two nitrogen atoms, N_2; the usual form in which nitrogen occurs in the Earth's atmosphere.

dopant Any material that is added in relatively small concentrations to a second material in order to change the properties of the second material.

drop tower A high tower from which objects are dropped to produce conditions of very low gravitational force.

drug enhancer Any substance that is added to or combined with a drug in order to increase its effect.

dubnium (Db) The name suggested by the International Union of Pure and Applied Chemistry for element number 105.

dye Any soluble material that is used to add color to some other material, as in a paint or ink.

emissions trading A practice commonly used in pollution control agreements whereby the savings on pollutants attained by one party can be traded to a second party as long as the net result of such an exchange represents an overall improvement in some environmental condition.

equimolar concentration Any solution in which the components are present in equal concentrations by mole.

erythrocyte A red blood cell.

eutectic The lowest melting point of an alloy or some other kind (generally) of metallic solution.

excitation energy   The amount of energy required to raise an electron in an atom from some lower state to some higher state or, more generally, to raise a particle from some lower energy state to a higher energy state.

face-centered cube   A type of crystal structure in which the unit cell contains an atom, ion, or other particle at each corner of a cube as well as at the center of each face of the cube.

femto-   A prefix standing for one-quadrillionth, 10^{-15}.

femtosecond   A quadrillionth (10^{-15}) of a second.

flame ball   A region of combustion that develops when a material burns in a zero-gravity or microgravity setting. Flame balls have properties that differ from the shapes of flames that develop in the presence of normal gravitational fields.

fluorescence   A form of luminescence in which the time that passes between the absorption of energy and the emission of light is very short (less than or equal to 10^{-3} second).

fossil fuel   Coal, oil, or natural gas.

fractal   A geometrical pattern that is self-repeating. Each portion of the pattern, when examined under higher magnification, exhibits the same pattern as at lower magnification.

free radical   A chemical species that contains one or more unpaired electrons and that is, therefore, generally very reactive.

genome   A complete set of chromosomes; it contains the sum total of all the genetic information of an organism or species.

graphene   A two-dimensional sheet of carbon atoms.

green fluorescent protein (GFP)   A form of protein found in the jellyfish *Aequoria victoria*, which lives off the coast of the Pacific Northwest. GFP is responsible for the greenish glow given off by these jellyfish in their natural habitat.

greenhouse gas   Any gas that can absorb solar energy re-radiated from the surface of Earth (in the form of infrared radiation), thus contributing to a warming of Earth's atmosphere.

halogenated aromatic hydrocarbon   An organic compound with a basic structure that includes one or more benzene-like rings in which all or some of the hydrogen atoms have been replaced by halogen atoms.

hassium (Hs)   The name suggested by the International Union of Pure and Applied Chemistry for element number 108.

heterocyclic ring   A chemical structure, usually organic, that contains two or more elements. A pyridine ring, for example, is heterocyclic because it consists of four carbon and one nitrogen atoms.

hot melt   A thermoplastic resin that melts when heated and then is used in that state to form a strong bond between materials upon cooling.

hydride   A hydrogen-containing binary compound in which hydrogen is the more negative element in the compound.

hydridic hydrogen   A weakly negative hydrogen ion.

hydrochlorofluorocarbons (HCFCs)   See *CFCs*. A HCFC differs from a CFC in having at least one hydrogen remaining in the final product.

hydrophilic   Having a tendency to bind to or be attracted to water.

hydrophobic   Having a tendency to be repelled by water.

in vivo   Literally, "in a living body." Commonly used to describe experiments carried out in living organisms, such as experimental animals or humans.

in vitro   Literally, "in glass." Generally used to describe experiments carried out outside a living organism, as in beakers, petri dishes, or other laboratory equipment.

isoprene   The common name for 3-methyl-1,3-butadiene $(CH_2=C(CH_3)CH=CH_2)$, the monomer from which natural rubber and gutta-percha are formed. Isoprene is also the fundamental unit present in terpenes.

isoprenoid   Any compound that contains the isoprene unit as a fundamental component of its structure.

latent pigment   A substance that can be converted into a pigment by means of some chemical and/or physical action.

ligand   An atom, ion, or molecule that is bonded to the central atom in a coordination compound, a chelate, or other type of chemical complex.

luminescence   The process by which a material absorbs energy and then re-emits it in the same or some other form. Some types of luminescence include bioluminescence, chemiluminescence, electroluminescence, photolumines-cence, fluorescence, and phosphorescence. The difference between the last two of these terms involves the time delay between the absorption and emission of energy by the material.

magic number   In nuclear chemistry, a combination of protons and neutrons in the nucleus of an atom that tends to provide unusual stability to that nucleus. The magic numbers include 2, 8, 20, 28, 50, and 82.

magnetic susceptibility   A measure of the amount to which a material is magnetized. Mathematically, it is the ratio of the magnetism of the material to the strength of the surrounding magnetic field.

meitnerium (Mt)   The name suggested by the International Union of Pure and Applied Chemistry for element number 109.

micelle   A charged colloidal particle. (A tiny droplet of oil surrounded by detergent molecules is an example of a micelle.)

microgravity   A condition of very low gravitational force.

mitochondria   Cell organelles found in eukaryotic cells. Their primary function is the production of energy through respiratory functions.

monomer   A chemical compound capable of undergoing polymerization. Adjectival form: monomeric.

nano-   The prefix meaning one-billionth (10^{-9}).

nanocrystal   A piece of crystal with dimensions of no more than a few nanometers.

nanotube   A structure consisting of carbon no more than a few atoms thick, but of substantially greater linear dimensions (a few micrometers, for example).

National Ambient Air Quality Standards (NAAQS)   Concentration limits set by the U.S. Environmental Protection Agency for six air pollutants—ozone, nitrogen oxides, sulfur dioxide, carbon monoxide, lead, and particulate matter—that are designed to protect human health.

neurochemistry   The study of the chemical structure and behavior of materials that make up nerve cells and the nervous system.

neurotransmitter   Any chemical that is able to carry a nerve signal from one neuron to an adjacent neuron.

noise   Any signal received from a system that does not provide useful information about that system.

nuclear magnetic resonance (NMR) spectroscopy   A mechanism for studying structural properties of matter by exposing it to a strong external magnetic field and measuring the changes produced by these fields.

oleophilic   Having a tendency to bond to or be attracted by lipids or other nonpolar materials.

oligomer   A molecule that consists of only a few (usually 2 to 10) amino acid residues bonded to each other.

ozone-depleting substance   Any chemical species that has the capability of reacting with and destroying ozone molecules.

peptide   A term that refers to the bonding between amino acids. It is sometimes used as a synonym for *polypeptide* and is often used to describe the bond between two amino acids, as in a *peptide bond.*

photoactivation   The process by which some material is raised to a higher energy level with light as, for example, in raising the energy level of electrons in an atom by shining light on the atom.

photochemical reaction   A chemical change that is brought about or modified by the application of radiant energy.

photolyase   An enzyme that is responsible for the repair of DNA molecules in a cell.

pico-   The abbreviation for one-trillionth (10^{-12}).

pigment   A substance that is used to color materials but usually is not soluble in solvents typically used for dyes and other coloring materials. (Exceptions are the natural organic pigments.)

plateau of stability   A section of the periodic table that contains one or more stable elements.

Poisson distribution   A mathematical function describing the likelihood that certain random events will occur in a given situation.

polyketide   A class of organic compounds that contain two or more carbonyl groups attached to a skeletal carbon chain.

polymer   A chemical compound of high molecular weight formed of one or two simple molecules (monomers) joined to each other in long chains. Adjectival form: polymeric.

polypeptide   A polymer consisting of many (usually at least ten, and often hundreds or thousands of) amino acids joined to each other by peptide bonds. Proteins are examples of polypeptides.

porous silicon   A form of silicon in which the material has been etched so that it has many grooves across its surface.

Portland cement   A form of cement containing aluminum, calcium, iron, and silicon oxide, with variable and complex composition.

precursor   A biochemical substance that existed in one form, usually inactive, that is then converted into a second form that has some biological or biochemical function. For example, carotene (pro-vitamin A) is a precursor of vitamin A because it is changed within the body into the active vitamin form.

primary structure   The sequence of amino acids that make up the polypeptide chain in a protein.

proenzyme   A protein precursor of an enzyme, that is, a molecule that exists in an inactive form in the body but that is converted by some chemical or physical process into an active form of the enzyme.

prostaglandin   A member of a group of biochemical compounds that mediate physiological functions. The prostaglandins are derived from arachidonic acid, which is found in the liver and glandular organs. Prostaglandins have a very wide range of biological effects.

protonation   The process by which a proton (or hydrogen ion) is added to a substance.

protonic hydrogen   A weakly positive hydrogen ion.

racemic mixture   A mixture of optically active isomers in which half rotate light one way and half rotate light the opposite way. Therefore, the mixture itself has no optical activity.

reductase   An enzyme that catalyzes reduction reactions.

replication (DNA)   The process by which a DNA molecule makes a copy of itself.

reverse micelle   A colloidal particle in which a tiny droplet of water is enclosed by a traditional micelle.

rhombohedral crystal structure   A crystal structure in which the unit cell is a parallelepiped whose faces are rhombuses. In other words, the unit cell is a rhombohedron.

rotamer   One of a variety of forms an amino acid residue may take in a protein, differing from other rotameric forms on the basis of its kinetic energy.

rutherfordium (Rf)   The name suggested by the International Union of Pure and Applied Chemistry for element number 104.

scanning tunneling microscope (STM) A type of microscope with a very sharp tip through which an electrical current is discharged. Electrical interactions between the tip and the surface being examined as the microscope is passed over a material provide a map of the material's surface.

seaborgium (Sg) The name suggested by the International Union of Pure and Applied Chemistry for element number 106.

self-replication The process by which a molecule (usually a nucleic acid or polypeptide) makes copies of itself.

sickle-cell anemia A genetic disorder in which a person's body produces red blood cells with distorted, usually sickle-like, shapes.

signal The set of useful information that can be obtained from a given system, compared to the noise, which is the set of all signals carrying no useful information from the system.

solvophilic Having an attraction to the molecules of a solvent. If the solvent is water, the more specific term is *hydrophilic*, while attraction to lipid solvents is described as *oleophilic*.

solvophobic Having a tendency to be repelled by the molecules of a solvent.

sonochemistry The study of chemical changes that result from bombardment by sound waves.

steric Having to do with the arrangements in space of the atoms in a molecule, ion, or other species.

substrate In general, any material on which some other material or force is deposited. One of the most common applications involves enzymes, where the term refers to the material on which an enzyme acts. Other common applications involve the etching of or deposition on a surface, in which case the term refers to the surface of the material being treated.

superconductor A material that is able to conduct an electrical current without resistance. Thus, once a current has been initiated in a superconductor, it theoretically continues to travel through the material forever.

supercritical fluid A gas at a temperature higher than its critical temperature and under high pressure, such that it displays the properties of both a gas and a liquid.

supramolecular complex A combination of (usually) molecules joined to each other in such a way that the grouping behaves like a single unit.

template In chemistry, a molecule that acts as a pattern for the construction of a complementary molecule.

teratogenic Anything having a tendency to produce malformed structures in a fetus or embryo.

terpene One of a large class of unsaturated hydrocarbons whose molecules contain two or more isoprene (C_5H_8) units.

tetrahedral structure of carbon atom The geometric arrangement formed in most cases by the four bonds in a carbon atom in organic compounds.

thalassemia   Any one of a group of genetic disorders that result in improperly formed hemoglobin molecules.

thiols   A class of organic compounds containing the -SH functional group.

transfermium element   An element with an atomic number greater than that of fermium (atomic number 100).

transition temperature   The temperature at which a substance becomes superconductive.

vitamin K cycle   The cycle in which vitamin K is first reduced to vitamin KH_2 by the action of the enzyme reductase, then converted to vitamin K oxide by the enzyme carboxylase, then converted back to vitamin K by a reductase enzyme.

X-ray diffraction analysis   The process by which crystal structure is determined by shining X-rays through the crystal. The interference pattern formed by the diffraction of the X-rays indicates the arrangement of atoms and ions within the crystal.

zinc finger   A naturally occurring polypeptide that is the smallest known polypeptide with the ability to fold in such a way as to include the three basic types of folding known in proteins.

zymogen   A proenzyme; that is, a molecule that is converted from an inactive form within the body into an active enzymatic form.

INDEX

AACCNews, 236
AATCC Technical Manual, 244
abiotic, 275
academia, chemical careers in, 206–07
acetylsalicylic acid, 275. *See also*
    aspirin
acoustic, 275
ACTION, 255
active block (in superconductors), 93,
    275
adenosine diphosphate (ADP), 51, 275
adenosine monophosphate (AMP), 51,
    276
adenosine triphosphate (ATP), 51, 276
Adhesive and Sealant Council, Inc.
    (ASC), 251
ADP. *See* adenosine diphosphate
*Advances in Fluid Particles and Fluid
    Particle Systems*, 240
Advances in Forest Product Industries,
    240
Aequoria victoria, 44
aging, 73–74
AIChE Journal, 240
AIChExtra, 240
Air Pollution Control Act of 1955, 126
Air Quality Act of 1967, 126

Albany Medical College, 76
Alberts, Bruce, 172–73
Albritton, Daniel L., 170–71
alkanes, 37
Alliance of Small Island States (ASIS),
    121–22
alpha helix, 53, 276
Alternative Agriculture, 176
American Association for Clinical
    Chemistry, Inc. (AACC), 236–37
American Association for the Advance-
    ment of Science (AAAS), 237
American Association of Cereal
    Chemists (AACC), 243
American Association of Pharmaceuti-
    cal Scientists (AAPS), 243
American Association of Textile
    Chemists and Colorists (AATCC),
    243–44
American Ceramic Society (ACerS),
    244
American Chemical Society (ACS),
    237–39
American Crop Protection Association
    (ACPA), 251
American Gas Association (AGA), 252
American Institute of Chemical
    Engineers (AIChE), 239–40

American Leather Chemists Association (ALCA), 245
American Oil Chemists' Society (AOCS), 245
American Petroleum Institute (API), 252
American Scientist, 259
American Society of Brewing Chemists (ASBC), 245
American Trucking Association, 128
AMP. *See* adenosine monophosphate
amino acid residue, 276
aminoboranes, 20–21
ammonia, 37
Ammonia Plant Safety and Related Facilities, 240
amphiphilic surfaces, 110–11, 276
analytical chemists, 209
Analytical Communications, 242
Andrews, Nancy C., 84, 136
antigen, 276
antiserum, 276
Antifungal Agents: Discovery and Modes of Action, 242
antigen, 276
antiserum, 276
AOAC INTERNATIONAL, 246
applied research, 205
arachidonic acid, 276
Armbruster, Peter, 137
The Armchair Scientist, 270
aromatic, 276
The ASC Journal, 251
aspirin, 65–67
ATP. *See* adenosine triphosphate
ATP synthase, 51–53
Aum Shinrikyo, 132

Baker, David, 54, 137
Baldeschwieler, John D., 201
Barbara, Paul, 27, 137–38
Bard, Allen J., 201, 203
Barton, Jacqueline K., 201, 202
basic research, 205
batteries, 105–06
Bednorz, J. Georg, 92
Benedick, Richard, 167
Berlin Mandate, 120
beta sheet, 53, 276
Biological Weapons Convention of 1972, 132–33, 192

bioprocess and biotechnology, future of, 200
biosensors, 98, 277
Biotechnology Progress, 240
Biotechnology Today, 243
bipolar particle, 277
blinking molecules, 27–28
blood-clotting cascade (BCC), 62–65, 277
blood transfusions, 75
blood types, 75
bohrium, 5, 277
Bojkov, Rumen, 169
bone marrow cancer, 28
Boyer, Paul D., 51, 138
Breslow, Ronald, 201
Brewer, Karen J., 96, 138
Brigham and Women's Hospital, 84
British Petroleum (BP), 122
Brown, Theodore L., 201
"bubble" policies, 121–22, 277
bubble-driven chemical reactions, 42–43
buckminsterfullerene, 12, 277
buckyballs, 12–15, 277
budget allocations for chemistry, U.S. federal, 228–29
bulk water, 104
Bush, George, 133
Business Alliance to Protect Americans from Chemical Weapons, 194
Byrd-Hagel Resolution, 121–22

Cady, John, 176–80
California Institute of Technology (Caltech), 58
Cambridge Crystallographic Database, 20
carbon dioxide, effect on climate, 118–19
carbon dioxide, supercritical. *See* supercritical carbon dioxide
carboxylation, 62–64
"Carcinogens and Anticarcinogens in the Human Diet," 177
Careers for Chemists: A World Outside the Lab, 217
careers in chemistry
    compensation patterns, 216
    educational requirements, 212–13

employment opportunities, 213–15
    resources, 216–20
Carnegie Institution Geophysical
    Laboratory, 11
catalysis, future research, 199–200
Catalyst, 251
cavitation, 43, 277
Center for High Pressure Research, 11
Center for Materials Science and
    Engineering (University of Texas),
    106
Center for Technological Education-
    Holon, 90
Ceramic Abstracts, 244
Ceramic Bulletin, 244
*Ceramic Engineering and Science
    Proceedings*, 244
"Ceramic Engineering: Career
    Opportunities for You," 217
ceramics, 97–98
ceramicSOURCE, 244
Cereal Chemistry, 243
Cereal Foods World, 243
CFCs, 113, 277
    atmospheric effects, 124–25
    disposal of, 113–15
chalcogen, 277
chaperone proteins, 68–69, 277
ChAPTER One, 240
charge-reserve block (in superconduc-
    tors), 93, 277
chelate, 277
Chem Matters, 261
ChemDex, 270
Chemical & Engineering News
    career articles and advertising
        supplements, 218
Chemical Communications, 242
chemical consultants, 208
chemical corporations, 226–27
The Chemical Educator, 260
Chemical Engineering, 270
chemical engineering, careers in, 210–
    11
Chemical Engineering Progress, 240
chemical industry, statistics and data,
    221–26
Chemical Information on the Internet,
    271
Chemical Manufacturers Association
    (CMA), 248–49

*Chemical Process Control: Assessment
    and New Directions for Research*,
    240
chemical products, statistics and data,
    223–24
    value of exports, 224
    value of imports, 223–24
Chemical Society Reviews, 242
Chemical Specialties Manufacturers
    Association (CSMA), 249–50
chemical synthesis, future of, 199–200
chemical technology, careers in, 211–
    12
Chemical Times & Trends, 249
chemical weapons, 131–32, 190–95,
    277
Chemical Weapons Convention, 131–
    34, 190–95
Chemical Week, 260
chemicals, statistics and data, 1990–95,
    221–22, 229–33
    value of shipments, 1990–95, 221–
    22
chemicals, toxic, 228
Chemistry & Industry, 242, 261
Chemistry and Industry Magazine, 243
Chemistry International, 241
Chemistry of Art, 265
chemistry teaching, 207
Chemistry Today, 243
"Chemistry's Golden Age," 201–03
chemists, statistics and data, 222–25
CHEMTECH, 262
Chemunity News, 262
Children's Hospital (Harvard Univer-
    sity), 82
chlorine, as chemical weapon, 132
Chlorine Chemistry Council (CCC),
    253
chlorofluorocarbons. *See* CFCs
Christianson, David W., 139
chromophore groups, 28, 45, 277
Chu, Ching-Wu (Paul), 92, 139–40
Ciba Specialty Chemicals, 94
clathrate, 277
Clean Air Acts, 117, 126–28
climate change, 118–23, 180–90
Clinical Chemistry, 236
Clinical Laboratory News, 236
Clinton, Bill, 121, 131, 133

coal, statistics and data, 233
Cole, Thomas W., Jr., 22
Collinson, Maryanne, 33
Color Pigments Manufacturers
    Association, Inc. (CPMA), 253
compensation for chemists, 223, 224,
    225
Composites Institute, 254
Comprehensive Environmental
    Response, Compensation, and
    Liability Act of 1980 (CERCLA),
    118.
*Consumer Products Education Bureau
    Quarterly Report*, 249
Consumer Topics in Focus, 249
Contemporary Organic Synthesis, 242
Copenhagen Amendments. *See*
    Montreal Protocol on Substances
    that Deplete the Ozone Layer
copper chaperones, 68–69
copper chips (semiconductors), 100–02
Cosmetic, Toiletry, and Fragrance
    Association (CTFA), 254
Council for Chemical Research (CCR),
    250
Crabtree, Robert H., 20–22, 140
Cracking It!, 216
Crop Protection Agents from Nature,
    242
Crutzen, Paul, 140–41
cubane, 22–24
Cummins, Christopher C., 40
curium, 3
Curl, Robert F., Jr., 12, 80, 141
cyclases, 48–49, 278
cyclization, 49–51, 278
cyclotetramethylenetetranitramine
    (HMX), 24

Dahiyat, Bassil I., 58, 141–42
dalton, 278
Dalton Transactions, 242
Dam, Carl Peter Nerik, 62
dead-end elimination (DEE) theorem,
    58
Dealer PROGRESS, 255
Delaney, James J., 129
Delaney clause, 129–31, 171–80
dendrimers, 15–20, 278
diamond anvil cell, 11, 36, 278

dihydrogen bonding, 20, 278
dinitrogen, 39, 41, 278
Discover, 262
Discover Magazine Virtual Bookstore,
disulfide bonds, 54
DNA synthesis, 69–71
dopant, 278
Dowd, Paul, 62, 142
Dow's Exposure Index, 240
drop towers, 107, 278
drug design, 85–88
dubnium, 5, 6, 278
Dufaux, Douglas P., 114
Dugan, Laura F., 142–43
dyes, 94, 278

earnings in the chemical industry, 223
Earth Summit. *See* United Nations
    Conference on Environment and
    Development
Eaton, Philip E., 22
Education in Chemistry, 263
Eindhoven Institute of Technology, 18,
    26
Einstein, Albert, 5
einsteinium, 5
The Electrochemical Society (ECS),
    246
Electrochemical Society Interface, 246
element 104. *See* rutherfordium
element 106. *See* seaborgium
Elf-Aquitane Corporation, 112
emissions trading, 120–21, 278
employment in chemistry and chemical
    engineering, 1990–95, 222–23,
    224
emulsification, 103–05
Environment Today, 243
environmental chemistry, careers in,
    211
Environmental Progress, 240
*Environmental Science & Technology
    Meetings*, 242
*Environmental Toxicology and
    Chemistry*, 248
enzymes, effect on protein repair, 73–
    74
erythrocytes, 75–77, 278
erythromycins, 77–80
Estrada-Oyuela, Raúl, 122

eutectic, 278
European Union (EU), 121–22
Executive Newsletter, 253
Executive Newswatch, 249
Exploring Animal Behavior, 260
Exploring Evolutionary Biology, 260

F_0F_1ATPase. *See* ATP synthase
Faraday Transactions, 242
Federal Food, Drug, and Cosmetic Act
    of 1958 (FFDCA), 129, 171, 178
Federal Insecticide, Fungicide, and
    Rodenticide Control Act of 1972
    (FIFRCA), 129, 178
femto, 279
Fermi, Enrico, 5
fermium, 5
Fertilizer Financial Facts, 255
The Fertilizer Institute (TFI), 254–55
Fertilizer Production Cost Surveys, 255
flame balls, 106–10, 279
flames, 106–10
fluorescent materials, 27, 279
food additives, 171–80
Food Quality Protection Act of 1996,
    130–31
Food Science & Technology Meetings,
    242
*Forensic Urine Drug Testing Newslet-
    ter*, 236
fossil fuels, 279
    combustion, 119
    statistics and data, 233–234
fractal, 279
free radical, 279
Fujishima, Akira, 110
Fuller, Buckminster, 12
fullerenes, 12, 80–82, 90–91

Garavito, R. Michael, 66, 143
Gehring, Peter M., 143–44
Geneva Protocol on Chemical Weap-
    ons, 132, 193
The Geochemical News, 247
The Geochemical Society (GS), 247
Geochimica et Cosmochimica Acta,
    247
Gesellschaft für Schwerionenforschung
    (GSI), 4
Ghadiri, M. Reza, 99, 144

global warming. *See* climate change
government, chemical careers in, 207–
    08
graphene sheets, 13, 279
Great Minds of Science, 263
green fluorescent protein (GFP), 44–
    48, 279
greenhouse effect, 118
greenhouse gases, 118–22, 279
Gulf War, 132

Haber-Bosch process, 40
halogenated aromatic hydrocarbons,
    112–13, 279
Handbook for Teaching Assistants, 265
Harvard Medical School, 84
hassium, 5, 279
Hazardous Materials Shipping Manual,
    251
Heat Transfer, 240
Hediger, Matthias A., 84
Helicobacter pylori, genome, 88–90
Helms, Jesse, 133–34
Hemley, Russell, J., 144–45
heterocyclic ring, 279
HIV, 28
HMX. *See* cyclotetramethylenetetrani-
    tramine
Hoeber, Francis P., 190–93
Hoffmann, Felix, 65
hopene, 49
hot melts, 31, 279
Howard Hughes Medical Institute, 58,
    60, 84
Howdle, Steven, 145
Hydra, 117
hydridic hydrogen, 21, 280
hydrogen, 35–37
hydrogen bonding, 8, 20–21, 26, 54,
    55
hydrogen bonding *(continued)*
    in polymers, 29–31
    hydrogen cyanide, as chemical
        weapon, 132

IBM, 101
Iijima, Sumio, 13, 145
Imperiali, Barbara, 201
in Chemistry, 263
Informatics in Chemistry, 271

information systems in chemistry,
    future of, 198
Inhofe, James M., 128
inorganic chemists, 209
Inside Laboratory Management, 246
Institut für Organishe Chemie and
    Biochemie, 49
Institute for Heavy-Ion Research. *See*
    Gesellschaft für
    Schwerionenforschung (GSI)
Institute for High-Pressure Physics
    (Russian Academy of Sciences), 11
Institute for Scientific Information,
    271–72
Institute of Optics and Quantum
    Electronics (Friedrich Schiller
    University), 41
International Union of Pure and
    Applied Chemistry (IUPAC), 5,
    241
*Internet Chemistry Resources: Guide to
    Internet Resources*, 272
*Internet Resources for Science and
    Mathematics Education*, 272
iron transportation in blood, 84–85
isoprenoids, 48–49, 280

John Innes Centre, 86
Johns Hopkins University, 68
Johnston, Keith P., 103
Joint Institute of Nuclear Research, 4
Journal of AOAC INTERNATIONAL,
    246
Journal of Chemical Education, 264–65
Journal of Chemical Research, 242
*Journal of Chemical Technology &
    Biotechnology*, 242
Journal of Materials Chemistry, 242
Journal of Materials Research, 247
*Journal of the American Ceramic
    Society*, 244
*Journal of the American Chemical
    Society*, 264
*Journal of the American Leather
    Chemists Association*, 245
*Journal of the American Oil Chemists'
    Society*, 245
Journal of the Electrochemical Society,
    246

*Journal of the Science of Food &
    Agriculture*, 242
Jupiter, 10, 32

Kaffia, 131
Kamerlingh-Onnes, Heike, 92
Keio University, 52
Khosla, Chaitan, 78, 145–46
Kids Discover, 265
Kim, Edward, 73
Kim, Jaekook, 106
Kleckner, Dean, 176–80
Kool, Eric T., 69, 146
Kroto, Harold W., 12, 80, 146–47
Kurchatov, Igor Vasilevich, 4
kurchatovium, 4
Kyoto Conference, 120–23
Kyoto Protocol to the United Nations
    Framework Convention on
    Climate Change, 180–90
Laboratorie d'Optique Appliquée, 41
Laboratorie de Chimie de Coordina-
    tion (CNRS), 18, 112
Laboratorie de Physique des Solids, 41
Laboratorie pour l'Utilisation des
    Lasers Intenses (at CNRS), 41
Langer, Robert S., 201
Laplaza, Catalina E., 40
Lawrence Berkeley National Labora-
    tory (LBNL), 4, 38
Lawrence Livermore National Labora-
    tory (LLNL), 9, 31
*The Learning Matter of Chemistry:
    Chemistry Information on the
    Internet*, 272
Lehmann, Paul, 169
LeToullec, René, 35
librarians, chemical, 208
ligand, 280
Lippard, Stephen, J., 201, 203
lithium batteries, 105–06
London Amendments. *See* Montreal
    Protocol on Substances that
    Deplete the Ozone Layer
Loubeyre, Paul, 35, 147
lubricants, solid, 90–91
luminescence, 280

magic numbers, 7, 280
Majoral, Jean-Pierre, 19, 147

manganese oxyiodide (in batteries),
    105–06
Manthiram, Arumugam, 106
manufacturing and operations in
    chemical technology, future of,
    198
marketing careers in chemistry, 206
Massachusetts Institute of Technology
    (MIT), 28, 40, 82
Materials Research Society (MRS), 247
materials technology, future of, 200–
    01
Mayo, S. L., 58, 148
Meijer, E. W., 18
meitnerium, 5, 280
Mendeleev Communications, 242
metal hydrides, 22
metal ion transport in cells, 67–69
metallic hydrogen, 8–11
methane clathrate hydrate, 31–32
methyl bromide, environmental effects,
    125–26
Meyerhoff, Albert H., 173–76
micelle, 280
microgravity, effect on flames, 107–10
minerals, industrial, statistics and data,
    229–32
mitochondria, 280
Modern Experiments for Introductory
    College Chemistry, 264
Molina, Mario J., 148–49
Montreal Protocol on Substances that
    Deplete the Ozone Layer, 124–25,
    166, 167, 170
Moore, Jeffrey S., 149–50
Motor Gasoline, 242
Motorola, 101
MRS Bulletin, 247
Müller, K. Alex, 90
mustard gas, 132

Nakanishi, Koji, 201
nanoparticles, as lubricants, 90–91
nanotubes, 13–15, 281
National Ambient Air Quality Stan-
    dards (NAAQS), 127–28, 281
National Coalition of Petroleum
    Retailers, 128
National Engineers Week, 240
National Environmental Policy Act of
    1969 (NEPA), 117

National Institutes of Standards and
    Technology, 23, 114
National Oceanic and Atmospheric
    Agency, chemical careers, 207
National Paint and Coatings Associa-
    tion (NPCA), 255
National Pest Control Association
    (NPCA), 255
natural gas, statistics and data, 234
Natural Products Report, 242
naturalSCIENCE, 272–73
Nature, 265
Nature Biotechnology, 265
Nature Genetics, 265
Nature Medicine, 265
Nature Structural Biology, 265
negligible risk standard (for food
    additives), 178–79
Nellis, William, 9, 150
neurotransmitter, 281
nerve cells
    damage, 82–84
    growth, 80–84
New Scientist, 265–66
Newman, Paul, 169
nitrogen fixation, 40
Nixon, Richard, 133
Nocera, Daniel G., 201, 202
Northwestern University, 68
Nutrition and Sport, 242

Occupational Outlook Handbook, 217
octanitrocubane, 24
Official Methods of Analysis of AOAC
    INTERNATIONAL, 246
O'Halloran, Thomas V., 150–51
oligomers, 56, 281
    folding in, 56–58
Only Connect, 261, 273
Oppenheimer, Michael, 167
Opportunities in Chemistry Careers,
    217
organic chemists, 209
oxygen, 35–37
ozone, 123
    health standards, 127–28
ozone hole, 123–24
ozone layer (stratosphere), 60, 123
    depletion of, 123–26, 163–71
Ozone Trends Panel, 167

pain suppression, molecular biology of,
    66–67
Park, Hee-Won, 151
Parkinson's disease, 28
particulates, health standards, 127–28
patent attorneys, chemical, 208
Perkin Transactions 1 and 2, 242
peptide, 281
Pesticide Science, 242
*Pesticides in the Diets of Infants and
    Children*, 175
petroleum, statistics and data, 234
Pharmaceuticals Today, 243
Phillips, George N., Jr., 151
photolyase, 60, 281
photosynthesis, 95–97
physical chemists, 209
pigments, 94–95, 281
planetary atmospheres, 12
plateau of stability, 7, 281
Poisson distribution, 33, 281
pollution control, methods of, 111–15
polyketides, 85–88, 282
Polymer International, 242
polymers. *See* supramolecular polymers
polypeptide, 282
Polyurethane Foam Association (PFA),
    256
Popular Science, 266
porous silicon, 98–100, 282
President's Newsletter, 258
primary structure (of proteins), 53, 282
*Principles of Cereal Science and
    Technology*, 243
Prinn, Ronald, 167
*Proceedings of the Ethylene Producers'
    Conference*, 240
Process Safety Progress, 240
product development, 205–06
Production Survey, 255
Proposition 65 (California), 175
prostaglandin H$_2$ synthase, 66–67
prostaglandins, 65–67, 282
protein design, 57–59
protein folding, 53–59
protonation, 282
protonic hydrogen, 21, 282
Pure and Applied Chemistry, 241
Pure Food and Drug Act of 1906, 129

Raber, Douglas, 201
racemic mixture, 282
Reaction Times, 266
Reagan, Ronald, 133
Recovery of Bioproducts, 242
red blood cells (RBCs). *See* erythro-
    cytes
reductase, 282
Remington, S. James, 44
Resources Conservation and Recovery
    Act of 1970 (RCRA), 117
Resource Recovery Act of 1970 (RRA),
    117
Responsible Care®, 248–49, 258
reverse micelles, 103–04, 282
Rh antigens, 75
Rice, Stuart A., 201
Rice University, 44
Rio Summit. *See* United Nations
    Conference on Environment and
    Development
Ronney, Paul D., 152
rotamers, 58, 282
Rousse, Antoine, 41, 152
Rowland, F. Sherwood, 153, 167
Royal Dutch Shell, 122
Royal Society of Chemistry (RSC),
    241–42
Rubber Manufacturers Association
    (RMA), 256
Russian Chemical Reviews, 242
Rutherford, Sir Ernest, 5
rutherfordium, 5, 282

Safety in the Chemical Laboratory, 264
salaries for chemists. *See* compensa-
    tion for chemists
sales careers in chemistry, 206
salt linkages (in proteins), 54
sarin, 132
Saturn, 10, 32
scanning tunneling microscopy (STM),
    29, 283
Schädel, Matthias, 154
Schmidt, Christine, 154
Schulz, Georg E., 154–55
SCI Bulletin, 242
Science, 237, 266
Science News, 267
Science NOW, 267, 273

The Sciences, 267–68
Science's Next Wave, 267, 273
Scientific American, 268
Scott, Mark D., 76, 155
Scripps Research Institute, 99
Seaborg, Glenn, 5
seaborgium, 6–8, 283
Sealant Specifications Compendium, 251
Secrets of Science, 263
Shen, Tsung-Chung, 29, 156
silicon biosensors, 98–100
Singer, S. Fred, 166–69
single-molecule spectroscopy (SMS), 27
Skaggs Institute for Chemical Biology, 99
Skou, Jens Christian, 156
Small Business Regulatory Enforcement Fairness Acts of 1996 (SBREFA), 128
Smalley, Richard E., 12, 14, 80, 157, 201, 203
The Soap and Detergent Association (SDA), 256–57
Society of Chemical Industry (SCI), 242–43
Society of Cosmetic Chemists (SCC), 247–48
Society of Environmental Toxicology and Chemistry (SETAC), 248
The Society of the Plastics Industry, Inc. (SPI), 257
SOFBALL. *See* Structure of Flame Balls at Low Lewis-number
Soil and Water Quality: An Agenda for Agriculture, 176
solvophilic, 283
solvophobic, 283
sonochemistry, 43, 283
squalene, 49
Stanford University, 78, 86
"starburst" dendrimers, 19
State University at Blacksburg (Virginia), 96
State University of New York at Buffalo, 103
Stern, Laura, 157
Structure of Flame Balls at Low Lewis-number (experiments), 108–10

sulfur, 11–12
The Sulphur Institute (TSI), 257–58
Sulphur Outlook, 258
superconductivity, 11–12, 92–94
supercritical carbon dioxide, 102–05
supercritical fluid, 283
superheavy elements. *See* transfermium elements
supply chain management, future of, 198
supramolecular polymers, 26–27, 283
supramolecular trimetallic complexes, 95–97
Suslick, Kenneth S., 43, 158
Synthetic Organic Chemical Manufacturers Association (SOCMA), 258

technical services careers in chemistry, 206
Technology Review, 268
Technology Vision 2020, 197–201
Tenne, Reshef, 158
teratogenic, 283
Texas Center for Superconductivity (University of Houston), 92
Textile Chemist and Colorist, 244
TFI/NACA Agri-Dealer Action Fax Network, 255
Third Session on the Conference of the Parties to the UN Framework Convention on Climate Change. *See* Kyoto Conference
Today's Chemist at Work, 269
Tokyo Institute of Technology, 52
Tomalia, Donald A., 159
TOTO, Ltd., 110
toxic chemicals. *See* chemicals, toxic
transfermium elements, 2–8, 284
    chemical properties, 5–6
    discovery, 4
    nomenclature, 4
    production, 3–4
transition temperature, 11, 92, 93
Transport Properties and Related Thermodynamics Data of Binary Mixtures, 240
trinitrotoluene (TNT), 24
tungsten disulfide, as a lubricant, 90–91
Twin Cities Newsletter, 251

Ube Research Laboratories, 97
ultraviolet radiation, 123, 165–66
    health effects, 60, 123
unemployment among chemists, 224
unidirectional solidification, 97
United Nations Conference on
    Environment and Development,
    120
United Nations Intergovernmental
    Panel on Climate Change, 167
Université Paris, 35
University of California at Berkeley, 38
University of California at Los Angeles,
    66, 73
University of California at San Fran-
    cisco, 73
University of Chicago, 23, 66
University of Colorado at Boulder, 103
University of Illinois at Urbana-
    Champaign, 29, 43, 56
University of Maryland, 23
University of Michigan, 68
University of Minnesota, 28
University of Nottingham, 103
University of Oregon, 44
University of Pittsburgh, 62
University of Rochester, 66, 69
University of Texas at Austin, 103
University of Texas Medical Center, 60
University of Tokyo, 110
University of Washington, 54
Uranus, 32
U.S. Chamber of Commerce, 128
U.S. Department of Agriculture,
    chemical careers, 207
U.S. Department of Energy, chemical
    careers, 207
U.S. Environmental Protection Agency
    (EPA), 127, 130
    chemical careers, 207
U.S. federal budget. See budget
    allocations for chemistry, U.S.
    federal
U.S. Food and Drug Administration
    (FDA), 130
    chemical careers, 207–08

U.S. Geological Survey (USGS), 31
UV-B radiation, 123. See also ultravio-
    let radiation

van der Waal's forces, 54, 55
Vane, John, 65
Vendors to the Trade, 249
Virginia Polytechnic Institute, 96
vitamin K, 62–65, 284
VOC/VOS Manual, 251

wages for chemists. See compensation
    for chemists
Waku, Yoshiharu, 97, 159
Walker, John E., 51, 159–60
Washington University School of
    Medicine (St. Louis), 80
Washington Wire, 255
Watson, Robert T., 164–66
Webber, Frederick L., 193–95
Weizmann Institute, 90
What's New in the World?, 273
Wightman, Robert M., 33, 160–61
Wigner, Eugene, 8
Wonder Science, 269
Working Chemists with Disabilities:
    Expanding Opportunities in
    Science, 216
The World-Wide Web Virtual Library:
    Chemistry, 273
X-ray diffraction, 41–42, 284

YAHOO!, 273–74
Yale University, 20
Yang, Fan, 44
Yildrim, Taner, 161
Yoshida, Masasuke, 161

Zachariah, Michael R., 114
Zambounis, J. S., 162
Zare, Richard, 201, 202
Zeld'ovich, Yakov B., 107
zero tolerance (in foods). See Delaney
    clause
zinc finger, 58–59, 284
zymogen, 284